HONDA 750 FOURS Owners Workshop Manual

by Jeff Clew

Member of the Guild of Motoring Writers

Models covered

Honda CB750. 736cc. UK January 1970 to June 1971
Honda CB750. 736cc. US June 1969 to 1970
Honda CB750 K. 736cc. UK June 1971 to May 1979
Honda CB750 K. 736cc. US September 1970 to 1978
Honda CB750 F. 736cc. UK October 1975 to May 1979
Honda CB750 F. 736cc. US October 1974 to 1978
Honda CB750 A Hondamatic. 736cc. US March 1976 to 1978

ISBN 978 0 85696 521 0

ABCDE
FGHI
3

Printed in India *(131-6S9)*

Haynes Group Limited
Sparkford, Yeovil,
Somerset BA22 7JJ, England

Haynes North America, Inc
2801 Townsgate Road,
Suite 340, Thousand Oaks,
CA 91361, USA

Disclaimer

There are risks associated with automotive repairs. The ability to make repairs depends on the individual's skill, experience and proper tools. Individuals should act with due care and acknowledge and assume the risk of performing automotive repairs.

The purpose of this manual is to provide comprehensive, useful and accessible automotive repair information, to help you get the best value from your vehicle. However, this manual is not a substitute for a professional certified technician or mechanic.

This repair manual is produced by a third party and is not associated with an individual vehicle manufacturer. If there is any doubt or discrepancy between this manual and the owner's manual or the factory service manual, please refer to the factory service manual or seek assistance from a professional certified technician or mechanic.

Even though we have prepared this manual with extreme care and every attempt is made to ensure that the information in this manual is correct, neither the publisher nor the author can accept responsibility for loss, damage or injury caused by any errors in, or omissions from, the information given.

Acknowledgements

We are indebted to Honda (UK) Limited for the technical assistance given so freely when this manual was being prepared. Fowlers of Bristol and Tippetts of Surbiton supplied the spare parts needed for the overhaul of the CB750 on which this manual is based. Brian Horsfall gave the necessary assistance with the overhaul and devised ingenious methods of overcoming the lack of service tools. Les Brazier took many of the photographs used and Tim Parker edited the text.

We also wish to acknowledge the help of Vincent and Jerrom Limited, Taunton, and Fran Ridewood and Company of Wells who provided much useful 'in depth' modification information, based on their experience as Honda agents. Illustrations and advice about tyre fitting were kindly supplied by the Avon Rubber Company.

The cover photograph was arranged through the courtesy of Mr. P. Marsh of Bower Hinton, Martock, Somerset.

About this manual

The author of this manual has the conviction that the only way in which a meaningful and easy-to-follow text can be written is to carry out the work himself, under conditions similar to those found in the average household. As a result, the hands seen in the photographs are those of the author. Even the machines are not new; examples which have covered a considerable mileage are selected, so that the conditions encountered would be typical of those encountered by the average rider/owner. Unless specially mentioned, and therefore considered essential, Honda service tools have not been used. There are invariably alternative means of slackening or removing some vital component when service tools are not available, but risk of damage is to be avoided at all costs.

Each of the seven Chapters is divided into numbered Sections. Within the Sections are numbered paragraphs. Cross-reference throughout the manual is quite straightforward and logical. For example, when reference is made 'See Section 6.2' it means Section 6, paragraph 2 in the same Chapter. If another Chapter were meant, the reference would read 'See Chapter 2, Section 6.2'. All photographs are captioned with a Section/paragraph number to which they refer, and are always relevant to the Chapter text adjacent.

Figure numbers (usually line illustrations) appear in numerical order, within a given Chapter. Fig. 1.1 therefore refers to the first figure in Chapter 1. Left hand and right hand descriptions of the machines and their component parts refer to the left and right when the rider is seated, facing forward.

Motorcycle manufacturers continually make changes to specifications and recommendations, and these, when notified, are incorporated into our manuals at the earliest opportunity.

We take great pride in the accuracy of information given in this manual, but motorcycle manufacturers make alterations and design changes during the production run of a particular motorcycle of which they do not inform us. No liability can be accepted by the authors or publishers for loss, damage or injury caused by any errors in, or omissions from, the information given.

Contents

Note: General descriptions and specifications are given in each Chapter immediately after the list of contents.
Fault diagnosis is given when applicable at the end of the Chapter.

Introduction to the Honda 4 750

Although for many years manufacturers have produced what are virtually replicas of the 'Works' racing machines, for sale to the general public, it is true to say that no machine captured the interest of the motor cyclist so vividly as the first production version of the Honda 4 cylinder. Already a legend in racing circles, the Honda 4 had represented a serious challenge whenever it appeared in International events. With riders such as the late Bob MacIntyre, Jim Redman and Mike Hailwood, the 4 demonstrated its supremacy on frequent occasions, irrespective of whether the 125 cc, 250 cc, 350 cc or 500 cc version was raced. Even the previously unbeaten multi-cylinder Italian models no longer had things their own way and were hard put to continue racing under truly competitive terms.

At the end of 1967 Honda withdrew from racing and commenced work on a road-going version of their in-line 4, scaled up to 750 cc. Without question it was designed to be the number one 'Superbike', a position it has occupied since its introduction in 1969. In engine layout it follows the lines of the racing machines closely, a feature heightened by the use of four separate carburettors and four sets of exhaust pipes and silencers, two on each side of the machine. A speedometer calibrated up to 150 mph and a tachometer with the red band commencing at 8500 rpm completed the 'street racer' effect, which led to such a peak of interest that over 61,000 Honda 750 cc 4's were sold IN THE USA ALONE in just over three years. In Britain, the 750 model was first imported during January 1970, designated the model CB 750. It enjoyed a similar reception from discerning motor cyclists and soon became a fairly familiar sight on the roads, even though most only saw its rapidly disappearing rear view. It is still in production today, substantially unchanged in outward appearance, although greatly improved as the result of an incredible number of modifications aimed at making a good product even better. The Honda Motor Company Limited has an uncanny knack of producing the right machine at the right time and establishing yet another trend for others to follow. The 750 4 is no exception.

Against a background of rumours of Honda's return to GP road racing, the CB 750F model has been introduced. It incorporates Honda's new styling, giving it the 'cafe racer' image. The four-into-one exhaust, and rear disc brake are the most obvious external changes. Other minor changes have been made also, bringing this popular model up to date.

EARLY HONDA 750

1976 HONDA 750F (SEE CHAPTER 7)

Ordering Spare Parts

When ordering spare parts for any Honda 750 4 it is advisable to deal direct with an official Honda agent, who should be able to supply most items ex-stock. Parts cannot be obtained from Honda (UK) Limited direct: all orders must be routed via an approved agent, even if the parts required are not held in stock.

Always quote the engine and frame numbers in full, particularly if parts are required for any of the earlier models.

The frame number is located on the left hand side of the steering head and the engine number is stamped on the upper crankcase, immediately to the rear of the two left hand cylinders.

Use only parts of genuine Honda manufacture. Pattern parts are available, some of which originate from Japan, but in many instances they may have an adverse effect on performance and/or reliability.

Honda do not operate a 'service exchange' scheme.

Some of the more expendable parts such as spark plugs, bulbs, tyres, oils and greases etc., can be obtained from accessory shops and motor factors, who have convenient opening hours, charge lower prices and can often be found not far from home. It is also possible to obtain parts on a Mail Order basis from a number of specialists who advertise regularly in the motor cycle magazines.

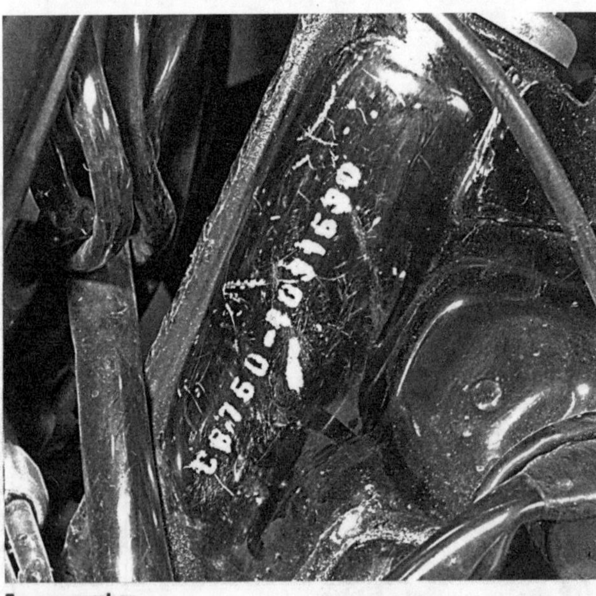

Frame number

Engine number

Dimensions and weights

	CB 750		CB 750F	
	in	mm	in	mm
Overall length	85.0	2160	86.6	2200
Overall width	34.8	885	33.9	860
Overall height	44.1	1120	45.7	1160
Wheelbase	57.3	1455	57.9	1470
Weight (dry)	480 lb	218 kg	499 lb	227 kg

Routine Maintenance

Periodic routine maintenance is a continuous process that commences immediately the machine is used. It must be carried out at specified mileage recordings or on a calendar date basis if the machine is not used regularly, at whichever falls soonest. Maintenance should be regarded as an insurance policy, to help keep the machine in peak condition and to ensure long, trouble-free service. It has the additional benefit of giving early warning of any faults that may develop and will act as a regular safety check, to the obvious advantage of both rider and machine alike.

The various maintenance tasks are described under their respective mileage and calendar headings. Accompanying diagrams are provided, where necessary. It should be remembered that the interval between the various maintenance tasks serves only as a guide. As the machine gets older or is used under particularly adverse conditions, it would be advisable to reduce the period between each check.

Some of the tasks are described in detail where they are not mentioned fully as a routine maintenance item in the text. If a specific item is mentioned, but not described in detail, it will be covered fully in the appropriate chapter. No special tools are required for the normal routine maintenance tasks. The tools contained in the tool kit supplied with every new machine will prove adequate for each task or if they are not available, the tools found in the average household will usually suffice.

Fortnightly or every 250 miles

Check level in oil tank and top up if necessary.
Check battery acid level and tyre pressures.
Check whether final drive chain is lubricated adequately and whether spring link is located securely (if fitted).

Monthly or every 500 miles

Check brake fluid level, check and if necessary adjust the rear brake pedal.
Service and adjust the final drive chain.
Check the clutch adjustment.
Examine both wheels for loose or damaged spokes.

RM1. Keep oil tank well topped-up

Check and if necessary adjust the valve clearances whilst the engine is cold; verify the accuracy of the ignition timing.
Adjust the camshaft drive chain.

Three monthly or every 2000 miles

Change the engine oil.

Six monthly or every 3000 miles

Clean and check all four spark plugs.
Check and if necessary service the contact breaker points.
Service the air cleaner, adjust the carburettors and verify they are synchronised correctly.
Check the petrol tap filter, petrol tank and fuel lines.
Check the front forks, paying particular attention to the security of the handlebars and the fork yokes.
Grease the swinging arm rear suspension pivot and check the

security of the rear suspension mounting bolts.
Check both wheel rims and hubs, the front brake caliper and the thickness of the pad linings. Do not omit to examine the rear brake linkage.
Make a visual check of the frame, oil tank and hoses, exhaust system and both stands.

Eight monthly or every 4000 miles

Change the oil filter element, drain and refill with new engine oil on the same occasion.

Yearly or every 6000 miles

Check the engine oil pressure.
Check the wiring system, especially the ignition circuit.
Check and if necessary adjust the steering head bearings, also the handlebar lock.
Change the oil in the front forks.
Check the front and rear wheel spindles and bearings, also the hydraulic brake lines. Remove and examine the rear brake linings.
Check the condition of the final drive sprockets, also that of the chain.

Two yearly or every 12,000 miles

Remove and clean the oil pump strainer.

It should be noted that even when six monthly and yearly maintenance tasks have to be undertaken, the fortnightly, monthly and three-monthly services must also be completed. There is no stage at any point in the life of the machine where a routine maintenance task can be ignored.

Mention has not been made of the lighting equipment, horn and speedometer, all of which must be checked frequently to ensure they are in good working order. The tyres also should be checked regularly and renewed if they wear unevenly, have splits or cracks in the side walls or if the depth of tread approaches the statutory minimum. Neglect of any of these points may render the owner liable to prosecution, apart from creating a safety hazard.

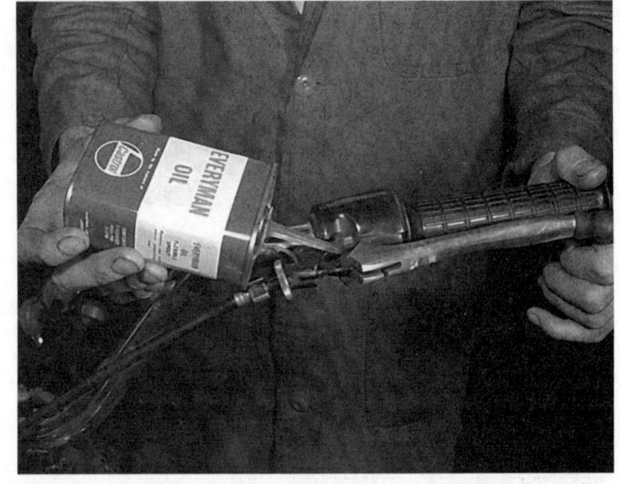

RM3. Do not overlook cables if smooth operation is to be maintained

RM4. Oil pump gauze must be cleaned every 12,000 miles

RM2. Never use anything other than the recommended brake fluid

RM5. Check oil level using the dipstick

Routine Maintenance and Capacities Data

Engine	
Oil tank capacity	6.2 Imp pints (7.4 US pints/3.5 litres)
Fuel tank capacity	3.7 Imp gallons (4.5 US gallons/17.0 litres)
Fuel tank reserve	1.1 Imp gallons (1.3 US gallons/5.0 litres)
Front forks - oil change	
CB750, K1 and K2	200 - 210 cc
CB750 K3 to K5	130 - 135 cc
CB750 K6	135 - 145 cc
CB750 F	125 - 135 cc
Valve clearances (engine cold)	
Inlet	0.002 in (0.05 mm)
Exhaust	0.003 in (0.08 mm)
Contact breaker gaps	0.012 - 0.016 in (0.3 - 0.4 mm)
Spark plug gaps	0.024 in - 0.028 in (0.6 - 0.7 mm)
Tyre pressures	
Front	28 psi *
Rear	28 psi *

*** For sustained high speed over 110 mph, increase to 32 psi and 34 psi respectively**

Recommended Lubricants

Component	Type of lubricant
Engine and gearbox	SAE 10W-40
Front forks	SAE 10W-30
All grease points, cam spindles, wheel hubs etc	Lithium-based high melting point grease
All lubrication points, control cables etc	Light oil, 10W-30 multigrade
Hydraulic disc brakes	High quality brake fluid

Safety first!

Professional motor mechanics are trained in safe working procedures. However enthusiastic you may be about getting on with the job in hand, do take the time to ensure that your safety is not put at risk. A moment's lack of attention can result in an accident, as can failure to observe certain elementary precautions.

There will always be new ways of having accidents, and the following points do not pretend to be a comprehensive list of all dangers; they are intended rather to make you aware of the risks and to encourage a safety-conscious approach to all work you carry out on your vehicle.

Essential DOs and DON'Ts

DON'T start the engine without first ascertaining that the transmission is in neutral.

DON'T suddenly remove the filler cap from a hot cooling system – cover it with a cloth and release the pressure gradually first, or you may get scalded by escaping coolant.

DON'T attempt to drain oil until you are sure it has cooled sufficiently to avoid scalding you.

DON'T grasp any part of the engine, exhaust or silencer without first ascertaining that it is sufficiently cool to avoid burning you.

DON'T allow brake fluid or antifreeze to contact the machine's paintwork or plastic components.

DON'T syphon toxic liquids such as fuel, brake fluid or antifreeze by mouth, or allow them to remain on your skin.

DON'T inhale dust – it may be injurious to health (see *Asbestos* heading).

DON'T allow any spilt oil or grease to remain on the floor – wipe it up straight away, before someone slips on it.

DON'T use ill-fitting spanners or other tools which may slip and cause injury.

DON'T attempt to lift a heavy component which may be beyond your capability – get assistance.

DON'T rush to finish a job, or take unverified short cuts.

DON'T allow children or animals in or around an unattended vehicle.

DON'T inflate a tyre to a pressure above the recommended maximum. Apart from overstressing the carcase and wheel rim, in extreme cases the tyre may blow off forcibly.

DO ensure that the machine is supported securely at all times. This is especially important when the machine is blocked up to aid wheel or fork removal.

DO take care when attempting to slacken a stubborn nut or bolt. It is generally better to pull on a spanner, rather than push, so that if slippage occurs you fall away from the machine rather than on to it.

DO wear eye protection when using power tools such as drill, sander, bench grinder etc.

DO use a barrier cream on your hands prior to undertaking dirty jobs – it will protect your skin from infection as well as making the dirt easier to remove afterwards; but make sure your hands aren't left slippery. Note that long-term contact with used engine oil can be a health hazard.

DO keep loose clothing (cuffs, tie etc) and long hair well out of the way of moving mechanical parts.

DO remove rings, wristwatch etc, before working on the vehicle – especially the electrical system.

DO keep your work area tidy – it is only too easy to fall over articles left lying around.

DO exercise caution when compressing springs for removal or installation. Ensure that the tension is applied and released in a controlled manner, using suitable tools which preclude the possibility of the spring escaping violently.

DO ensure that any lifting tackle used has a safe working load rating adequate for the job.

DO get someone to check periodically that all is well, when working alone on the vehicle.

DO carry out work in a logical sequence and check that everything is correctly assembled and tightened afterwards.

DO remember that your vehicle's safety affects that of yourself and others. If in doubt on any point, get specialist advice.

IF, in spite of following these precautions, you are unfortunate enough to injure yourself, seek medical attention as soon as possible.

Asbestos

Certain friction, insulating, sealing, and other products – such as brake linings, clutch linings, gaskets, etc – contain asbestos. *Extreme care must be taken to avoid inhalation of dust from such products since it is hazardous to health.* If in doubt, assume that they *do* contain asbestos.

Fire

Remember at all times that petrol (gasoline) is highly flammable. Never smoke, or have any kind of naked flame around, when working on the vehicle. But the risk does not end there – a spark caused by an electrical short-circuit, by two metal surfaces contacting each other, by careless use of tools, or even by static electricity built up in your body under certain conditions, can ignite petrol vapour, which in a confined space is highly explosive.

Always disconnect the battery earth (ground) terminal before working on any part of the fuel or electrical system, and never risk spilling fuel on to a hot engine or exhaust.

It is recommended that a fire extinguisher of a type suitable for fuel and electrical fires is kept handy in the garage or workplace at all times. Never try to extinguish a fuel or electrical fire with water.

Note: *Any reference to a 'torch' appearing in this manual should always be taken to mean a hand-held battery-operated electric lamp or flashlight. It does* **not** *mean a welding/gas torch or blowlamp.*

Fumes

Certain fumes are highly toxic and can quickly cause unconsciousness and even death if inhaled to any extent. Petrol (gasoline) vapour comes into this category, as do the vapours from certain solvents such as trichloroethylene. Any draining or pouring of such volatile fluids should be done in a well ventilated area.

When using cleaning fluids and solvents, read the instructions carefully. Never use materials from unmarked containers – they may give off poisonous vapours.

Never run the engine of a motor vehicle in an enclosed space such as a garage. Exhaust fumes contain carbon monoxide which is extremely poisonous; if you need to run the engine, always do so in the open air or at least have the rear of the vehicle outside the workplace.

The battery

Never cause a spark, or allow a naked light, near the vehicle's battery. It will normally be giving off a certain amount of hydrogen gas, which is highly explosive.

Always disconnect the battery earth (ground) terminal before working on the fuel or electrical systems.

If possible, loosen the filler plugs or cover when charging the battery from an external source. Do not charge at an excessive rate or the battery may burst.

Take care when topping up and when carrying the battery. The acid electrolyte, even when diluted, is very corrosive and should not be allowed to contact the eyes or skin.

If you ever need to prepare electrolyte yourself, always add the acid slowly to the water, and never the other way round. Protect against splashes by wearing rubber gloves and goggles.

Mains electricity and electrical equipment

When using an electric power tool, inspection light etc, always ensure that the appliance is correctly connected to its plug and that, where necessary, it is properly earthed (grounded). Do not use such appliances in damp conditions and, again, beware of creating a spark or applying excessive heat in the vicinity of fuel or fuel vapour. Also ensure that the appliances meet the relevant national safety standards.

Ignition HT voltage

A severe electric shock can result from touching certain parts of the ignition system, such as the HT leads, when the engine is running or being cranked, particularly if components are damp or the insulation is defective. Where an electronic ignition system is fitted, the HT voltage is much higher and could prove fatal.

Chapter 1 Engine, Clutch and Gearbox

Contents

Specifications

Engine

Type	Four cylinder, overhead camshaft, four stroke, air cooled
Cylinder arrangement	In line, mounted transversely across frame
Bore	61 mm (2.401 in)
Stroke	63 mm (2.480 in)
Capacity	736 cc (44.93 cu in)
Compression ratio	9.0 : 1
Cylinder compression	150 - 170 psi (10.5 - 12 kg sq cm)
Bhp	67 @ 8000 rpm

Cylinder head

Valve guide internal diameter (both valves)	0.2599 - 0.2603 in (6.6 - 6.61 mm)
Stem to guide clearance	
Inlet valves	0.003 in (0.08 mm) maximum (limit)
Exhaust valves	0.004 in (0.1 mm) maximum (limit)
Valve spring free length	
Inner springs	1.50 in (38.1 mm)
Limit of wear	1.4566 in (37.0 mm)
Outer springs	1.6220 in (41.2 mm)
Limit of wear	1.4748 in (40.0 mm)

Valve timing	
Inlet opens	5° BTDC
Inlet closes	30° ATDC
Exhaust opens	35° BTDC
Exhaust closes	5° ATDC
Rocker arm spindles	
Clearance between supports and spindles	0.005 in (0.12 mm) maximum (limit)
Rocker arm spindle diameter	0.4711 - 0.4718 in (11.966 - 11.984 mm)
Limit of wear	0.4701 in (11.94 mm)
Rocker arm bearing diameter	0.4724 - 0.4731 in (12.00 - 12.018 mm)
Limit of wear	0.4744 in (12.05 mm)

Camshaft

Cam height	
Inlet cams	1.4111 in (35.86 mm)
Exhaust cams	1.392 in (35.36 mm)
Base circle diameter	1.1016 - 1.1030 in (27.98 - 28.02 mm)
Limit of wear	1.099 in (27.93 mm)
Camshaft bearings diameter	0.8669 - 0.8678 in (22.02 - 22.04 mm)
Limit of wear	0.8701 in (22.0 mm)
Camshaft run-out	0.004 in (0.1 mm) maximum

Cylinders, pistons and rings

Cylinder bore diameter	2.4022 - 2.4024 in (61.01 - 61.02 mm)
Limit of wear	2.4055 in (61.1 mm)
Taper and ovality (maximum)	0.002 in (0.05 mm)

Pistons

Diameter	2.4002 - 2.4009 in (60.965 - 60.985 mm)
Limit of wear	2.3957 in (60.85 mm)
Gudgeon pin diameter	0.5903 - 0.5906 in (14.994 - 15.000 mm)
Limit of wear	0.589 in (14.96 mm)

Piston rings

End gap (compression)	0.008 in (0.2 mm)
Limit of wear	0.027 in (0.7 mm)
End gap (oil control)	0.004 in (0.1 mm)
Limit of wear	0.027 in (0.7 mm)
Side clearance (compression rings)	0.007 in (0.18 mm) maximum
Side clearance (oil control ring)	0.005 in (0.12 mm) maximum

Crankshaft and connecting rods

Main bearing and big end journals	Selective fits. See Section 1.16
Limit of wear (main bearings)	1.415 in (35.94 mm)
Taper and ovality (main bearings)	0.002 in (0.05 mm) maximum
Crankpins - diameter	1.4169 - 1.4173 in (35.99 - 36.00 mm)
Limit of wear	1.415 in (35.94 mm)

Clutch

Thickness of friction plates	0.1347 - 0.1409 in (3.42 - 3.58 mm)
Limit of wear	0.122 in (3.1 mm)
Plate warpage	0.012 in (0.3 mm) maximum
Clutch springs - free length	1.2575 in (31.94 mm)
Limit of wear	1.201 in (30.5 mm)

Torque wrench settings

Camshaft bearing cap nuts	7 - 9 lb ft
Connecting rod cap nuts	15 lb ft
Cylinder head cover screws	6 - 8 lb ft
Crankcase bolts (6 mm)	8 lb ft
Crankcase bolts (8 mm)	17 - 18 lb ft
Cylinder head bolts (6 mm)	8 lb ft
Cylinder head nuts	14 - 15 lb ft
Oil filter bolt	20 - 24 lb ft
Clutch locknut	33 - 36 lb ft
Rocker arm spindle bolts	6 - 8 lb ft

1 General description

The in-line four cylinder air-cooled engine fitted to the Honda CB 750 models is a combined engine and gearbox unit, in which the crankcase castings form the enclosure for the gearbox and final drive reduction gear. Aluminium alloy is used for all the engine and gearbox castings, giving efficient cooling and achieving a significant saving in weight.

The cylinder head carries the chain driven overhead camshaft which is driven through a tunnel between the two middle cylinders. The combustion chambers are semi-spherical in design; the valve guides are shrink fitted. The big ends and main bearings are of the shell type, the latter supporting the one-piece built-up crankshaft assembly. Only the gearbox has ball journal bearings.

Primary transmission consists of two short heavy duty primary chains, running in close proximity to each other, from twin sprockets in the centre of the crankshaft assembly to twin sprockets on the end of the gearbox mainshaft. The gearbox has five speeds and transmits the drive to the rear wheel by a secondary chain and sprockets, via a reduction gear contained within the gearbox casting.

The camshaft chain, which runs from the centre of the crankshaft assembly, is adjustable for wear by means of a chain tensioner which adjusts itself quite automatically when necessary. There is no provision for adjusting the primary chains. When wear develops, renewal of both chains is essential.

2 Operations with engine/gearbox in frame

It is not necessary to remove the engine/gearbox unit from the frame if only ancillary items such as the clutch, alternator and gear change mechanism require attention. The height and overall dimensions of the engine preclude the possibility of partially stripping the engine whilst it is still in the frame in the event of a cylinder head overhaul or similar repair work. The engine must first be lifted out as a complete unit and dismantled on the workbench.

3 Operations with engine/gearbox removed

1 Removal and replacement of the cylinder head and cylinder block.
2 Removal and replacement of the main bearings and big ends.
3 Removal and replacement of the crankshaft assembly.
4 Removal and replacement of the overhead camshaft.
5 Removal and replacement of the twin primary chain drive.
6 Removal and replacement of the gearbox components, selectors, kickstarter mechanism and reduction gear.

4 Method of engine/gearbox removal

As described previously, the engine and gearbox are built in unit and it is necessary to remove the unit complete in order to gain access to either unit. Separation of the crankcases is achieved after the engine unit has been removed from the frame and refitting cannot take place until the engine/gear unit is assembled completely. Access to the gearbox is not available until the engine has been dismantled and vice-versa in the case of attention to the bottom end of the engine. Fortunately, the task is made easy by arranging the crankcases to separate horizontally.

5 Removing the engine/gearbox unit

1 Place the machine on the centre stand, so that it is standing firmly on level ground. Place a receptacle that will hold at least a gallon under the crankcase and remove the drain plug so that the oil will drain off. It is preferable to do this whilst the engine is warm, so that the oil will drain more readily.
2 Transfer the receptacle containing the oil drainings to the

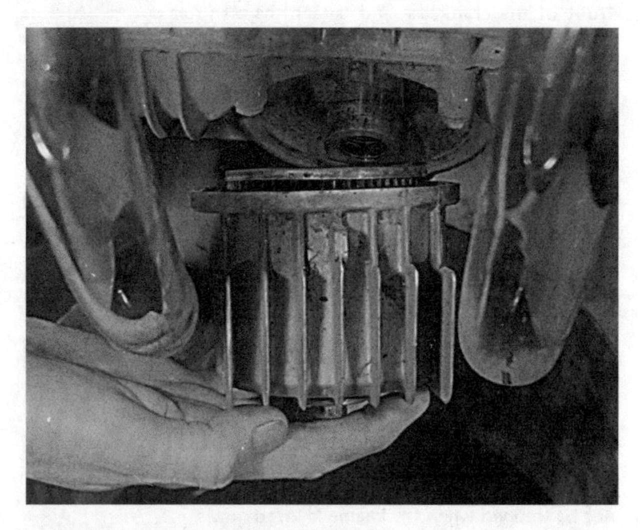

5.2. Oil will drain as filter is removed

5.3. Rubber cross-strap retains rear of tank

5.4. Disconnect positive wire from battery

front of the crankcase and detach the oil filter cover, which is retained by a single forward facing bolt. The cover will lift off, releasing a further quantity of oil and the oil filter element. Do not lose the O ring which seals the joint between the cover and the crankcase casting.

3 Raise the seat and pull the rubber cross strap from the lip at the rear of the petrol tank with which it engages. Close the petrol tap, release the retaining clips securing each of the twin fuel lines; pull off the fuel lines. The petrol tank can now be lifted away from the machine and stored in a place of safety, away from naked lights.

4 Whilst the seat is raised, disconnect the negative lead from the battery to isolate the electrical system and prevent accidental short circuiting.

5 Disconnect the tachometer drive cable from the right hand side of the cylinder head cover. It is retained by a single cross-head screw which must be slackened and detached first. The cable will pull from the fixing.

6 Check that each spark plug cap is numbered before removing them, otherwise difficulty will be encountered on reassembly. Tape the leads and caps to a nearby frame tube, so that they will not be trapped when the engine is lifted out.

7 On early models fitted with a throttle cable to each carburettor, unscrew each carburettor top and lift out the throttle valve, return spring and needle as a complete assembly. Tape each assembly to a conveniently placed frame tube; the slide and needle are very easily damaged if this precaution is not observed. Slacken off the clamp on the flexible inlet stub, nearest to each carburettor and the clamp which retains each carburettor intake to the air cleaner box. Each series of clamps has a cross-head screw. It will be necessary to prise the intake stub from each carburettor because a lipped fitting is used. When the carburettors are free, they can be withdrawn as a complete in-line unit, still attached to their mounting plate.

8 Later models have a push-pull cable arrangement and it is necessary to detach both cables from the link lever before the carburettors can be freed as described in the preceding paragraph. Before the carburettor tops can be released and the throttle valve assembly removed, it is necessary to unscrew the two 6 mm screws so that each individual carburettor can be detached from the mounting plate.

9 Remove the air filter box complete with the filter element. This is attached to the frame by two bolts at the top of the casing. If carburettor removal proves difficult, these bolts can be withdrawn first, to permit the air cleaner box to be moved backwards and give a small amount of extra clearance.

10 Detach each exhaust system by unscrewing the bolt through each finned exhaust pipe clamp. The exhaust pipes and silencers are manufactured as an integral unit and it is necessary to withdraw the bolt through each pillion footrest so that the silencers are released from their attachment to the frame. Note how a boss on the underside of each upper silencer engages with a rubber buffer attached to the upper surface of the lower silencer. Each complete exhaust system will now pull off the stub in the cylinder head.

11 The cylinder head stubs must be detached since there is insufficient clearance to remove the engine from the frame whilst they are still in situ. As the engine is lifted out from the right hand side, there is no necessity to remove the extreme right hand stub which is already outside the frame tubes. Each stub is retained by two cross head screws, to which a sealant has been applied. The use of an impact screwdriver is essential for this operation, using extreme care to prevent the screws from shearing as they emerge.

12 Remove both footrests which are bolted direct to the frame. Take off the kickstarter crank, which is retained on its splined shaft by a pinch bolt, and the rear brake pedal which is attached in a similar manner. It will be necessary first to detach the spring which actuates the stop lamp switch in this latter case.

13 Detach the circular, chromium-plated clutch cover from the right hand side of the engine unit. It is retained by three cross-head screws. Access is now available to the clutch operating arm, which can be turned anticlockwise to relax the tension on the

5.5. Screw retains tachometer drive take-off

5.7. Throttle slides are easily damaged

5.7a. Withdraw carburettors complete with mounting plate

5.10. Release exhaust systems at port end first

5.10a. Pillion footrests act also as silencer mounting bolts

5.10b. Rubber buffers surround silencer interconnections

5.12. Footrests bolt through frame gussets

5.13. Use spanner to release tension on clutch cable

5.13a. Withdraw cable stop to release cable

clutch cable. Detach the cable nipple from the end of the operating arm before the pressure is released, then withdraw the cable completely by unscrewing the cable stop from the outside of the clutch housing.

14 At this stage it is convenient to drain the oil tank by pulling off the right hand side cover and unscrewing the hexagon-headed drain plug from the bottom of the tank. The tank holds just over six pints of oil and a receptacle of the appropriate size should be positioned below.

15 Whilst the contents of the oil tank are draining off, remove the gear change lever from the left hand side of the machine. It is retained on its splined shaft by a single pinch bolt. Mark the lever and the shaft prior to removal, so that the lever is replaced at the same angle.

16 Detach the cover over the gearbox final drive sprocket, which is retained by two cross head screws. There is sufficient clearance for the cover to be lifted away without need to disturb the larger cover over the gear change mechanism. If the final drive chain is fitted with a spring link (early models) this can be removed and the chain separated so that it can be lifted away. If the chain is of the endless type, a chain rivet extractor will be needed to effect the separation. Either task is made easier if the section of the chain involved is positioned on the rear wheel sprocket.

17 When the oil tank has drained completely, detach the two large diameter oil hoses from their joints below the clutch housing. Each is retained by two bolts. More oil will be released during this operation, since the hoses have not drained. Detach and lift away the oil tank complete with hoses; it is retained by three bolts.

18 Pull off the left hand side cover and detach the starter motor cable from the starter solenoid switch. This is the thick cable, easily recognisable. Detach the electrical connections at the junction box in front of the battery and the lead wire to the stop lamp switch. Pull off the breather pipe from the rear of the cylinder head cover, after first releasing the wire clip.

19 Support the engine unit from the underside and withdraw the engine mounting bolts. A long bolt passes through lugs welded to the lower frame tubes, on the underside of the engine, and another through the upper rear portion of the crankcase casting. The remainder are short bolts, some of which pass through small, detachable engine plates fitted to the front and rear of the engine unit on the right hand side only. When all the bolts and engine plates have been removed, the engine is ready for lifting out of the frame from the right hand side, raising the rear end to clear the right hand gusset plate welded to the frame tubes.

20 Although the engine unit gives the appearance of being heavy on account of its overall bulk, it can be lifted out of the frame and transferred to the workbench quite easily with the aid of a second person. Stout metal rods inserted through the mounting lugs in the crankcase castings can be used to guide the unit out of the frame until a full lift can be applied.

6 Dismantling the engine/gearbox - general

1 Before commencing work on the engine, the external surfaces must be cleaned thoroughly. A motor cycle engine has very little protection from road dirt which will sooner or later find its way into the dismantled engine if this simple precaution is not observed.

2 One of the proprietary engine cleaning compounds such as Gunk or Jizer can be used to good effect, especially if the compound is allowed to penetrate the film of oil and grease before it is washed away. When washing down, make sure that water cannot enter the carburettors or the electrical system, particularly if these parts are now more exposed.

3 Never use force to remove any stubborn part, unless mention is made of this requirement in the text. There is invariably good reason why a part is difficult to remove, often because the dismantling operation has been tackled in the wrong sequence.

4 Dismantling will be made easier if a simple engine stand is constructed that will correspond with the engine mounting points. This arrangement will permit the complete unit to be

5.16. Final drive sprocket has small protective cover

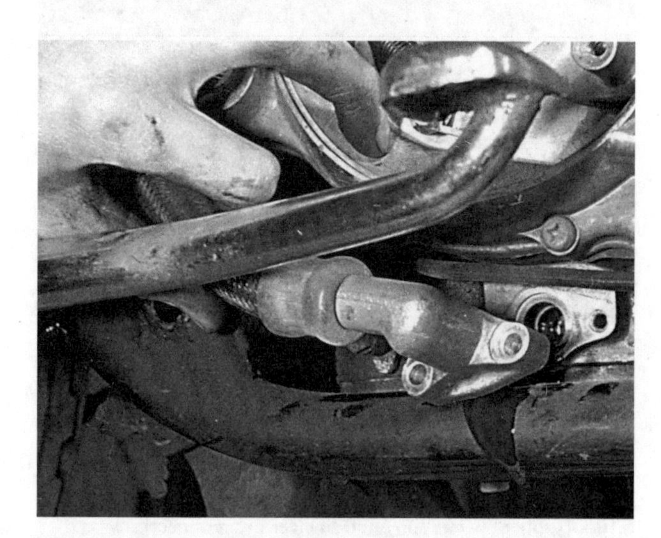

5.17. Detach oil hoses from crankcase

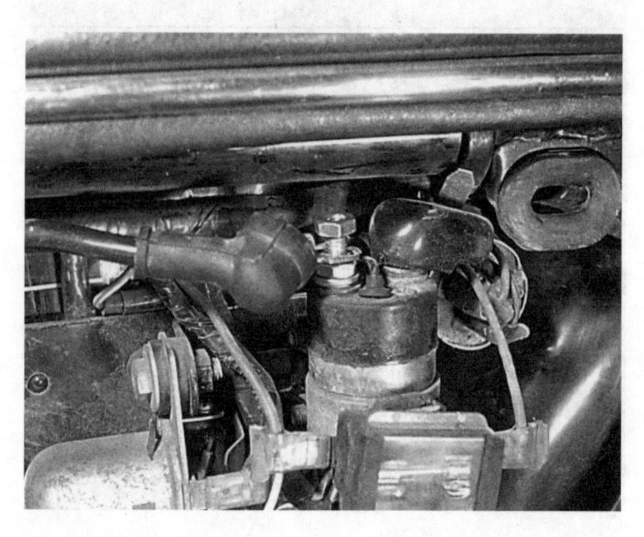

5.18. Disconnect starter motor cable at solenoid switch

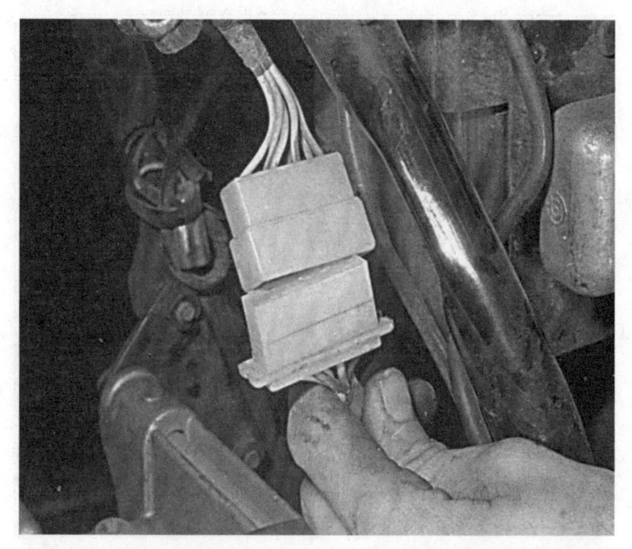

5.18a. Break electrical connections at junction boxes

5.19. Upper rear engine bolt makes earth connection

5.19a. Long engine bolt passes through underside of engine unit

5.19b. Right hand side of engine uses short engine plates at front..

5.19c. ... and rear of frame

5.20. Lift engine unit out from right-hand side

7.1. Breather cover masks two cylinder head cover screws

7.2. Align camshaft as shown before dismantling engine

7.3. Detach chain tensioner to release slack in chain

7.4. Lower bolt has just sufficient clearance for release

7.4a. Remove bolts to release rocker arm shafts

7.5. Lift endless chain off camshaft sprocket

7.5a. Sprocket cutaways enable sprocket to be pulled off camshaft

7.6. Camshaft holders lift off cylinder head studs

8.1. Screws are recessed deep into cylinder head

8.1a. Some nuts are located close to spark plugs

8.1b. Do not overlook the two small bolts at front and rear of head

8.4. Separate guide roller by withdrawing pin in centre

clamped rigidly to the workbench, leaving both hands free for the dismantling operation.

7 Dismantling the engine/gearbox unit - removing the cylinder head cover, rocker gear and overhead camshaft

1 Unscrew the three crosshead screws which retain the breather cover to the cylinder head cover. Slacken and remove the 16 crosshead screws of the cylinder head cover and lift the cover away.
2 Align the camshaft timing mark on the right hand end of the camshaft so that it is parallel with the cylinder head joint and has the keyway uppermost. Slacken off the caps which secure the camshaft bearings.
3 Detach the camshaft chain tensioner holder from the rear of the cylinder block by withdrawing the three bolts. The tensioner rod and spring will be released with the holder.
4 Unscrew the two bolts which retain the camshaft sprocket to the camshaft. There is just sufficient room for the bolts to be withdrawn. Loosen off all the valve clearance adjusting screws, remove the four bolts which lock the rocker arm shafts in position and pull out each shaft. Mark each rocker arm before it is released so that it is replaced in the same position during reassembly.
5 Remove the endless camshaft drive chain from the sprocket and pull the camshaft from its holder on the left hand side of the cylinder head, after removing the bearing caps. The internal cutaways in the camshaft sprocket will enable the sprocket to slide over the cams and be withdrawn from the camshaft; the endless chain can then be slid along the camshaft whilst the latter is withdrawn completely from the right hand side. Use a length of welding rod which will pass through the chain, to prevent it from dropping into the crankcase.
6 Lift off the two camshaft bearing holders by sliding them up the retaining studs which project from the cylinder head.

8 Dismantling the engine/gearbox unit - removing the cylinder head and cylinder block

1 The cylinder head is held in position by a combination of sixteen nuts and five bolts, with an additional four crosshead screws recessed into the top of the cylinder head beneath rubber inserts. Note that some nuts are recessed into the cutaways in the vicinity of the spark plugs. The correct sequence for slackening and subsequent tightening must be observed if distortion of the cylinder head is to be avoided. See accompanying diagram.
2 Lift off the cylinder head by raising it upward along the holding down studs whilst passing the camshaft drive chain through the tunnel between the two middle cylinders.
3 The cylinder block can then be raised in similar fashion. Care is necessary to support each piston as it emerges from the cylinder bore, to avoid damage.
4 The camshaft chain tensioner can now be removed from the top of the crankcase. To separate the guide roller and its mounting plate, push out the pin which passes through the centre of the roller.

9 Dismantling the engine/gearbox unit - removing the pistons and piston rings

1 Remove both circlips from each piston boss and discard them. Circlips should never be re-used if risk of displacement is to be obviated.
2 Using a drift of the correct diameter, tap each gudgeon pin out of the piston bosses until the piston can be lifted off the connecting rod, complete with rings. Make sure the piston is supported during this operation, or there is risk of bending the connecting rod.
3 If the gudgeon pin is a tight fit, do not resort to force. Warm the piston by placing a rag soaked in hot water on the crown, so that the piston bosses will expand and release their grip on the pin.
4 As each piston is removed, mark it with the cylinder number inside the skirt. There is no necessity to mark the back and front because this is denoted by an arrow cast in the crown, which must face towards the front of the machine on reassembly.
5 Each piston is fitted with three piston rings, two compression and one oil scraper. To remove the rings, spread the ends sufficiently with the thumbs to allow each ring to be eased from its groove and lifted clear of the piston. This is a very delicate operation necessitating great care. Piston rings are very brittle and will break easily.

10 Dismantling the engine/gearbox unit - removing the alternator and gear change mechanism

1 Remove the covers from the left hand side of the engine unit. The right hand cover is secured by a total of nine crosshead screws and the left hand cover by eight similar screws, around the periphery of each. Do NOT remove the three crosshead screws within the circular plate in the centre of the alternator cover. They secure the alternator stator coils and in no way impede movement of the cover complete. Remove the right hand cover first.
2 Separate the leads from the stator coil assembly at the snap connector joints, and place the left hand cover complete with the coil assembly in a safe place until reassembly commences.
3 Lock the engine by placing a stout metal rod through the small end of the extreme right hand connecting rod and remove the bolt from the centre of the alternator rotor. With the engine still locked in position, draw the alternator rotor off the tapered end of the crankshaft using either Honda Service tool 07011 - 30001, or a three-legged sprocket puller. The rotor will most probably pull off the shaft with the starter motor free running clutch pinion attached to the rear.
4 Remove the starter motor reduction gear from the shaft on which it is a light push fit, then pull out the shaft, using a pair of long nosed pliers.
5 Withdraw the gear change arm, of which the gear pedal spindle is an integral part, the side plate from the end of the gear selector drum (retained by one centre bolt), the gear selector drum stopper and the gear change positive stop plate. These latter two are secured by a stud and a bolt respectively. Make special note of the manner in which these components are assembled and the position of their return springs. Finally, remove the stud and the locking plate which secure the gear selector shaft.

11 Dismantling the engine/gearbox unit - removal of the contact breaker assembly and clutch

1 Working from the right hand side of the engine unit, remove the circular contact breaker cover, which is retained by two crosshead screws. Remove the 6 mm nut in the centre of the contact breaker assembly and then lift off the hexagon shaped washer which locates with projections on the end of the shaft carrying the contact breaker cam.
2 Unscrew the three crosshead screws around the periphery of the contact breaker base plate and lift away the complete contact breaker assembly with the electrical connections attached. The automatic advance unit will then pull from the long stud which threads into the end of the crankshaft. This stud is a tight fit and will shear if handled roughly. There is no need to withdraw the stud since it in no way impedes removal of the crankshaft.
3 Remove the clutch outer cover, held by ten crosshead screws around the periphery, thus providing access to the clutch. Unscrew its four retaining bolts and withdraw the clutch lifter plate with the four clutch springs.
4 Flatten back the tab locking the clutch centre nut. If the Honda peg spanner, Part Number 07096 - 300001, is not avail-

10.1. Rear left-hand cover encloses gear change mechanism

10.1a. Forward cover encloses alternator

10.2. Alternator wires separate at snap connectors

10.3. Use puller to draw alternator off tapered shaft

10.4. Pull out starter motor reduction gears

10.5. Withdraw gear change arm by lifting arm off selector drum

10.5a. Take end cover off selector drum, then ...

10.5b. ... dismantle stopper arms, before ...

10.5c. ... positive stop plate can be released

10.5d. Stud has locking plate for selector drum

11.1. Unscrew nut and pull off hexagonal washer ...

11.2. ... before contact breaker base plate can be removed

11.2a. Automatic advance unit will pull off centre stud

11.3. Remove clutch outer cover to gain access to the clutch...

11.3a. ... and unbolt the lifter plate

11.4. Use of welding rod to unscrew sleeve nut ...

11.4a. ... then lift out the clutch plates

11.5 Remove washer and circlip (later models), then use sprocket puller to pull off clutch outer drum

able to slacken the nut an alternative is to use a short length of welding rod bent into the shape of a flat U so that it can be pressed into two of the opposing slots. If a socket spanner of suitable size is then pressed into place on the bridged nut, it can be unscrewed without risk of damage. Select top gear and apply the rear brake to lock the transmission, then unscrew the nut and remove the tab washer and spring washer behind it. Withdraw as a single unit the clutch centre, f ,ction and plain plates and the pressure plate.

5 Remove the washer from the mainshaft and on later models, remove the 40 mm circlip from the end of the primary drive sleeve. Pull the outer drum off the sleeve by inserting the legs of a sprocket puller through the holes in the backplate. A splined fitting is employed, similar to that of the inner drum but larger in diameter. The fit is usually much tighter. The internal threads of the shaft MUST BE PROTECTED by inserting a bolt of the correct thread, when a puller is used.

6 Reverting to the left hand side of the engine, remove the four countersunk crosshead screws securing the layshaft main bearing holder in position. The plate will come away complete with the ball journal bearing.

11.6. Unscrew and remove layshaft bearing holder

12 Dismantling the engine/gearbox unit - removing the starter motor

1 The starter motor is located below the oblong chromium plated lid immediately to the rear of the cylinder block. To remove the lid, withdraw the two retaining bolts.

2 The starter motor has a splined shaft which engages directly with the intermediate reduction pinion. If the two bolts which pass through the right hand end of the starter motor casting are removed, it can be lifted up from that end and withdrawn from its recessed housing with the electrical cable attached. Note that it is advisable to disconnect the neutral indicator lead from the neutral contact first, since this lead passes through the starter motor housing. It is attached to the neutral contact by a small crosshead screw which is accessible when the protective cover is prised back.

3 It should be emphasised that there is no necessity to strip the engine or even remove the engine from the frame if the starter motor has to receive attention whilst the machine is in service. The procedure described in this section is sufficient; even the crankcase side covers need not be disturbed.

12.2. Raise starter motor upwards at rear to release from housing

13 Separating the crankcases

1 Slacken and remove the upper crankcase securing bolts. There is a total of eleven; eight 6 mm bolts, two 8 mm bolts and one 10 mm bolt.

2 Invert the crankcase assembly and remove the bolts from the lower crankcase. There is a total of twenty bolts and two nuts, comprising ten 6 mm bolts, ten 8 mm bolts, one 8 mm nut and one 10 mm nut.

3 The crankcases can now be separated, preferably by inverting them again so that there is less risk of damaging the long holding down studs. Tap around the jointing area with a rawhide mallet and on no account use a screwdriver to force the joints apart since this will cause irreparable damage to the mating surfaces.

4 Note that there is no necessity to remove the finned baseplate, if the oil pump does not require attention. It is retained by its own separate bolts which, if left in position, do not impede crankcase separation. For information relating to access to the oil pump and the dismantling procedure necessary, refer to Chapter 2.

14 Dismantling the gear clusters and selectors and removing the crankshaft

1 Remove the locking bolt from the oil retaining plate which retains the reduction gear assembly and lift out the reduction gear complete with final drive sprocket.

2 Unscrew the selector fork shaft retainer and pull out the

14.1. Reduction gear is retained by oil retainer

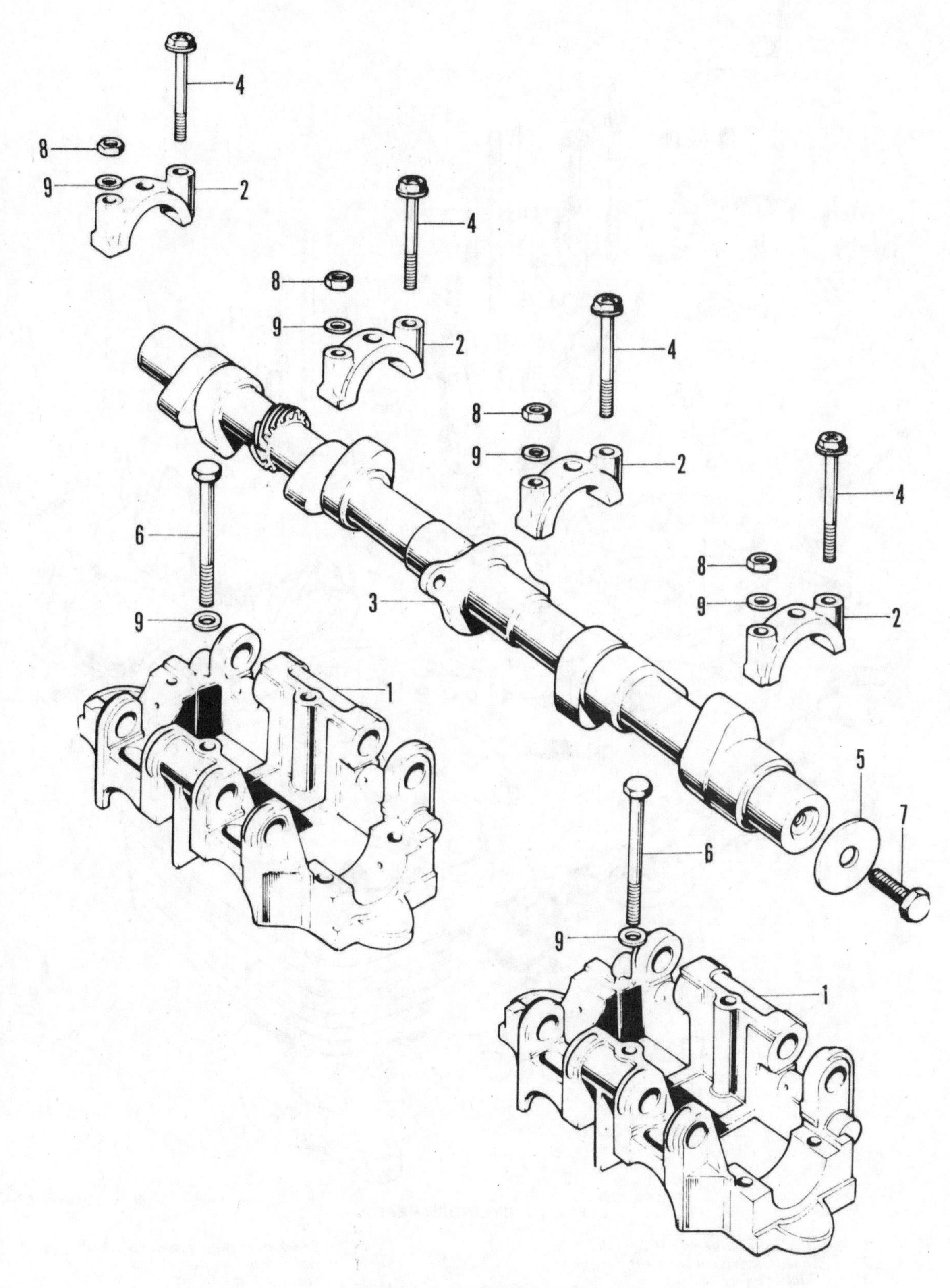

FIG. 1.1. OVERHEAD CAMSHAFT

1	*Camshaft holder - 2 off*	*6*	*Bolt - 4 off*
2	*Camshaft holder cap - 2 off*	*7*	*Bolt*
3	*Camshaft*	*8*	*Nut - 4 off*
4	*Bolt - 4 off*	*9*	*Plain washer - 8 off*
5	*Thrust washer*		

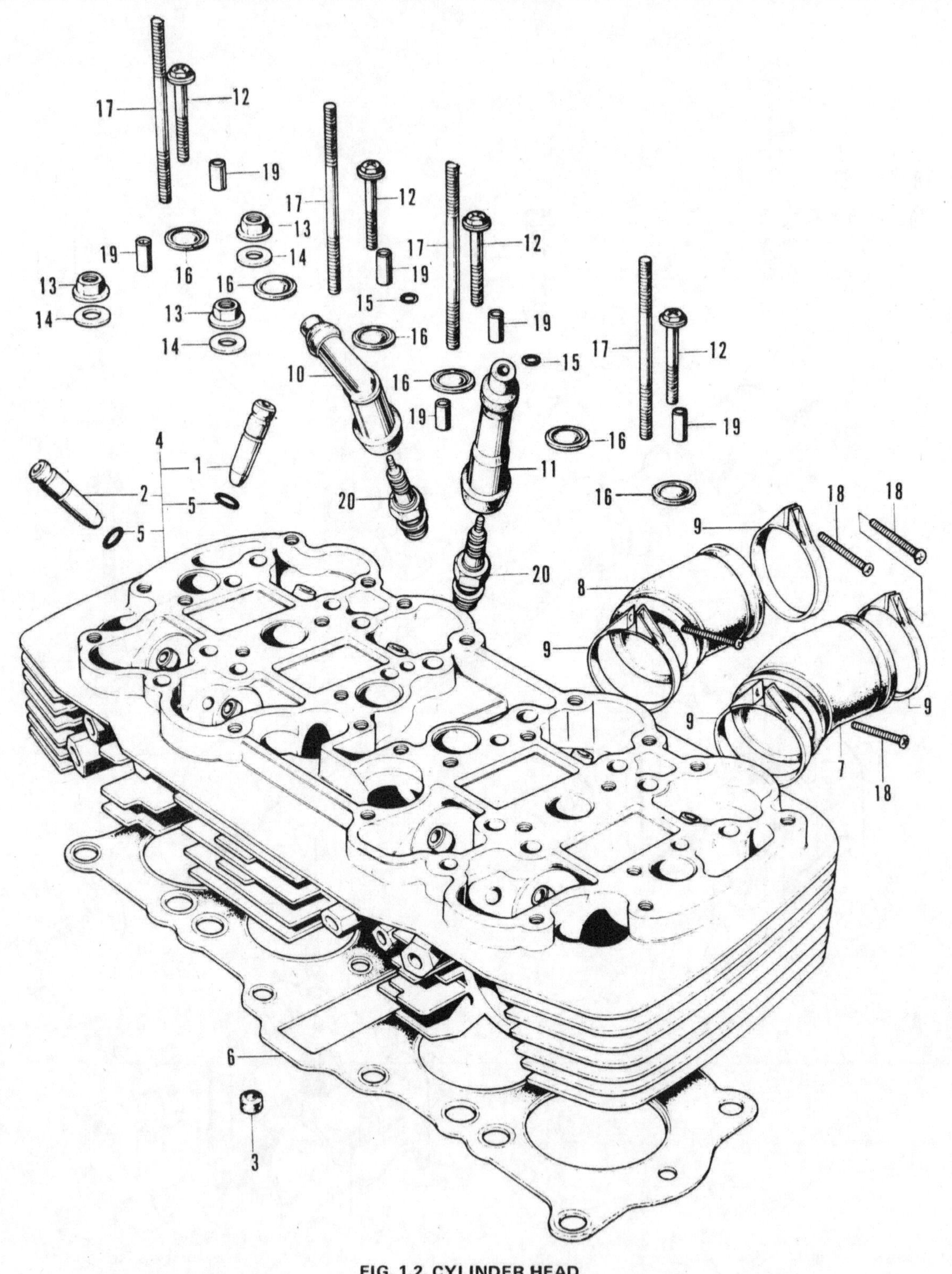

FIG. 1.2. CYLINDER HEAD

1	*Inlet valve guide - 4 off*	11	*Spark plug suppressor cap - 2 off*
2	*Exhaust valve guide - 4 off*	12	*Bolt - 5 off*
3	*Rubber insert - 12 off*	13	*Nut - 8 mm - 16 off*
4	*Cylinder head casting*	14	*Washer - 14 mm - 16 off*
5	*Valve guide clip - 8 off*	15	*'O' ring - 2 off*
6	*Cylinder head gasket*	16	*Rubber seals - 6 off*
7	*Heat insulator A - 2 off*	17	*Stud - 4 off*
8	*Heat insulator B - 2 off*	18	*Cross head screw - 8 off*
9	*Clamp - 8 off*	19	*Dowel pin - 6 off*
10	*Spark plug suppressor cap - 2 off*	20	*Spark plug - 4 off*

selector shaft from the left hand side of the engine unit by screwing an engine bolt of the correct size into the end and using this as an extractor. Lift out the middle of the three selector forks.

3 Lift the gearbox mainshaft at the left hand end and withdraw the mainshaft from the splined centre of the primary drive sprocket. The dual primary chains can then be lifted off the sprocket and the sprocket removed, complete with the right hand main bearing and internal needle roller bearing.

4 Lift out the two remaining selector forks, move the layshaft to the left so that the extreme left hand pinion will pass through the aperture in the lower crankcase casting and lift out the final drive pinion, followed by the layshaft bottom gear pinion, both of which will be released as the layshaft is withdrawn from their centres. Pull off the layshaft top gear pinion, which is now outside the crankcase casting and lift the layshaft away, complete with the three remaining gear pinions.

5 The kickstarter pinion and ratchet assembly is retained by a slotted pin which locks the splined kickstarter shaft in position. Withdraw the pin with a pair of thin nosed pliers, then pull the shaft from the centre of the pinion. The pinion will then lift out of position, permitting the return spring to be detached.

6 Lift the complete crankshaft assembly from the lower crankcase. The dual primary drive chains and the camshaft drive chain are of the endless type, but can be removed quite easily by lifting them off along the crankshaft.

7 Remove the primary chain tensioner, which is retained by a single bolt, the various bearing shells (marking each to ensure correct relocation), the split rings which locate the gearbox ball journal bearings and the large diameter hollow dowel and O ring which forms the main oil feed connection between the upper and lower crankcases. From the underside of the lower crankcase remove its retaining bolt and stop plate and withdraw the neutral contact, the selector drum can then be withdrawn to the left. With the exception of the oil pump assembly, the crankcases are now stripped completely.

15 Examination and renovation - general

1 Before examining the parts of the dismantled engine unit for wear, it is essential that they should be cleaned thoroughly. Use a paraffin/petrol mix to remove all traces of old oil and sludge that may have accumulated within the engine.

2 Examine the crankcase castings for cracks or other signs of damage. If a crack is discovered, it will require professional repair.

3 Carefully examine each part to determine the extent of wear, checking with the tolerance figures listed in the Specifications section of this Chapter. If there is any question of doubt, play safe and renew.

4 Use a clean, lint-free rag for cleaning and drying the various components. This will obviate the risk of small particles obstructing the internal oilways, causing the lubrication system to fail.

16 Big ends and main bearings - examination and renovation

1 Bearing shells are relatively inexpensive and it is prudent to renew the entire set of main bearing shells when the engine is dismantled completely, especially in view of the amount of work which will be necessary at a later date if any of the bearings fail. Always renew the five sets of main bearings together.

2 Wear is usually evident in the form of scuffing or score marks in the bearing surface. It is not possible to polish these marks out in view of the very soft nature of the bearing surface and the increased clearance that will result. If wear of this nature is detected, the crankshaft must be checked for ovality as described in the following section.

3 Failure of the big-end bearings is invariably accompanied by a pronounced knock within the crankcase. The knock will become progressively worse and vibration will also be experienced. It is essential that bearing failure is attended to without delay because if the engine is used in this condition there is a risk of breaking

Big end bearing – Metal selection table

	Connecting rod			Crank pin			Bearing			
Marking	1	2	3	3	4	5	Black	Brown	Green	Yellow
Dimension (mm)	39.000 to 39.008	39.008 to 39.016	39.016 to 39.024	36.000 to 35.995	35.995 to 35.990	35.990 to 35.985	15.002 to 14.998	14.998 to 14.994	14.994 to 14.990	14.900 to 14.986
Oil clearance										
20–41		1			3		Yellow (13218–300–013)			
22–43		1			5		Green (13217–300–013)			
25–46		1			4		Yellow (13218–300–013)			
20–41		2			3		Green (13217–300–013)			
22–43		2			4		Green (13217–300–013)			
25–46		2			5		Brown (13216–300–013)			
20–41		3			3		Brown (13216–300–013)			
22–43		3			4		Brown (13216–300–013)			
25–46		3			3		Black (13215–300–013)			

Main bearing – Metal selection table

	Crankcase			Crankshaft			Bearing			
Marking	A	B	C	A	B	C	Black	Brown	Green	Yellow
Dimension (mm)	39.000 to 39.008	39.008 to 39.016	39.016 to 39.024	36.000 to 35.995	35.995 to 35.990	35.990 to 35.985	15.002 to 14.993	14.998 to 14.994	14.994 to 14.990	14.990 to 14.986
Oil clearance										
20–41		A			A		Yellow (13318–300–013)			
22–43		A			C		Green (13317–300–013)			
25–46		A			B		Yellow (13318–300–013)			
20–41		B			A		Green (13317–300–013)			
22–43		B			B		Green (13317–300–013)			
25–46		B			C		Brown (13316–300–013)			
20–41		C			A		Brown (13316–300–013)			
22–43		C			B		Brown (13316–300–013)			
25–46		C			C		Black (13315–300–013)			

a connecting rod or even the crankshaft, causing more extensive damage.

4 Before the big-end bearings can be examined the bearing caps must be removed from each connecting rod. Each cap is retained by two high tension bolts. Before removal, mark each cap in relation to its connecting rod so that it may be replaced correctly. As with the main bearings, wear will be evident as scuffing or scoring and the bearing shells must be renewed as four complete sets.

5 New big-end shells are selected according to the size markings on the crankshaft and connecting rod, and main bearing shells according to the size marks on the crankcase upper half and crankshaft. Refer to the accompanying table to identify the bearing colour code required. Both big-end and main bearing shells are colour-coded with a dab of paint on one edge.

6 When selecting big-end shells, refer to the number stamped across each connecting rod and cap (1, 2 or 3) and the numbers (3, 4 or 5) stamped on the crankshaft web. Those on the crankshaft web represent the crankpin size and will be preceded by a circled P, the letter L denotes that they are to be read from the left side of the crankshaft.

7 In the case of the main bearings, refer to the letters stamped in the upper crankcase half, on the engine mounting extension, and the letters stamped on the crankshaft web; in both cases the letters A, B or C will be found. A circled J precedes the markings on the crankshaft, and the L indicates that they are read from the left side.

17 Crankshaft assembly - examination and renovation

1 If wear has necessitated the renewal of the big end and/or main bearing shells, the crankshaft bearing surfaces must be checked with a micrometer to verify whether ovality has occurred. If the reading varies by more than 0.002 inch, the crankshaft must be renewed.

2 Mount the crankshaft by supporting both ends on V blocks and check the run-out at the centre bearing surface by means of a dial guage. The amount of run-out will be half that of the gauge reading indicated. If the run-out is greater than 0.002 inch, specialist attention is needed to straighten the crankshaft.

3 When refitting the connecting rods and shell bearings, note that under no circumstances should the shells be adjusted with a shim, scraped in or the fit 'corrected' by filing the connecting rod and connecting rod cap or by applying emery cloth to the bearing surface. Treatment such as this will end in disaster; if the fit is not good, the parts concerned have not been assembled correctly. This advice also applies to the main bearing shells.

4 Oil the bearing surfaces before reassembly takes place and make sure the end tags of the bearing shells are located correctly. After the initial tightening of the connecting rod nuts, check that each connecting rod revolves quite freely, then tighten with a torque wrench to a setting of 14.5 lb ft (2 kg m). Check again that the bearing is still free.

18 Connecting rods - examination and renovation

1 It is unlikely that any of the connecting rods will bend during normal usage, unless an unusual occurrence such as a dropped valve has caused the engine to lock. Carelessness when removing a tight gudgeon pin can also give rise to a similar problem. It is not advisable to straighten a bent connecting rod; renewal is the only satisfactory solution.

2 The small end eye of the connecting rod is unbushed and it will be necessary to renew the connecting rod if the gudgeon pin becomes a slack fit. Always check that the oil hole in the small end eye is not blocked since if the oil supply is cut off, the bearing surfaces will wear very rapidly.

19 Oil seals - examination and replacement

1 An oil seal is fitted to the right hand end of the crankshaft assembly, to prevent oil from entering the contact breaker. There is also an oil seal fitted behind the gearbox final drive sprocket. If either seal is damaged or has shown a tendency to leak, it must be renewed.

2 Oil seals also tend to lose their effectiveness if they harden with age. It is difficult to give any firm recommendations in this respect except to say that if there is any doubt about the condition of a seal, renew it as a precaution.

20 Cylinder block - examination and renovation

1 The usual indication of badly worn cylinder bores and pistons is excessive smoking from the exhausts and piston slap, a metallic rattle that occurs when there is little or no load on the engine. If the top of the bore of the cylinder block is examined carefully, it will be found that there is a ridge on the thrust side, the depth of which will vary according to the rate of wear that has taken place. This marks the limit of travel of the uppermost piston ring.

2 Measure the bore diameter just below the ridge, using an internal micrometer. Compare this reading with the diameter at the bottom of the cylinder bore, which has not been subjected to wear. If the difference in readings exceeds 0.12 mm (0.005 in) the cylinder should be rebored and fitted with an oversize piston and rings.

3 If an internal micrometer is not available, the amount of cylinder bore wear can be measured by inserting the piston without rings so that it is approximately ¾ inch from the top of the bore. If it is possible to insert a 0.005 inch feeler gauge between the piston and the cylinder wall on the thrust side of the piston, remedial action must be taken.

4 Honda can supply pistons in four oversizes: 0.25 mm (0.010 inch), 0.050 mm (0.020 inch), 0.075 mm (0.030 inch) and 1.0 mm (0.040 inch). The 1 mm oversize is the safe limit to which the cylinder bores can be increased in diameter.

5 Check that the surface of the cylinder bores is free from score marks or other damage that may have resulted from an earlier engine seizure or a displaced gudgeon pin. A rebore will be necessary to remove any deep scores, irrespective of the amount of bore wear that has taken place, otherwise a compression leak will occur.

6 Make sure the external cooling fins of the cylinder block are not clogged with oil or road dirt, which will prevent the free flow of air and cause the engine to overheat.

21 Pistons and piston rings - examination and renovation

1 Attention to the pistons and piston rings can be overlooked if a rebore is necessary, since new components will be fitted.

2 If a rebore is not considered necessary, examine each piston closely. Reject pistons that are scored or badly discoloured as the result of exhaust gases by-passing the rings.

3 Remove all carbon from the piston crowns, using a blunt scraper, which will not damage the surface of the piston. Clean away all carbon deposits from the valve cutaways and finish off with metal polish so that a clean, shining surface is achieved. Carbon will not adhere so readily to a polished surface.

4 Check that the gudgeon pin bosses are not worn or the circlip grooves damaged. Check that the piston ring grooves are not enlarged. Slide float should not exceed 0.007 inch (0.18 mm) compression rings or 0.005 inch (0.12 mm) oil control rings.

5 Piston ring wear can be measured by inserting the rings in the bore from the top and pushing them down with the base of the piston so that they are square in the bore and about 1½ inches down. If the end gap exceeds 0.027 inch (0.7 mm) on all rings, renewal is necessary.

6 Check that there is no build up of carbon on the inside surface of the rings or in the grooves of the pistons. Any build-up should be removed by careful scraping.

7 The piston crowns will show whether the engine has been rebored on some previous occasion. All oversize pistons have the rebore size stamped on the crown. This information is essential when ordering replacement piston rings.

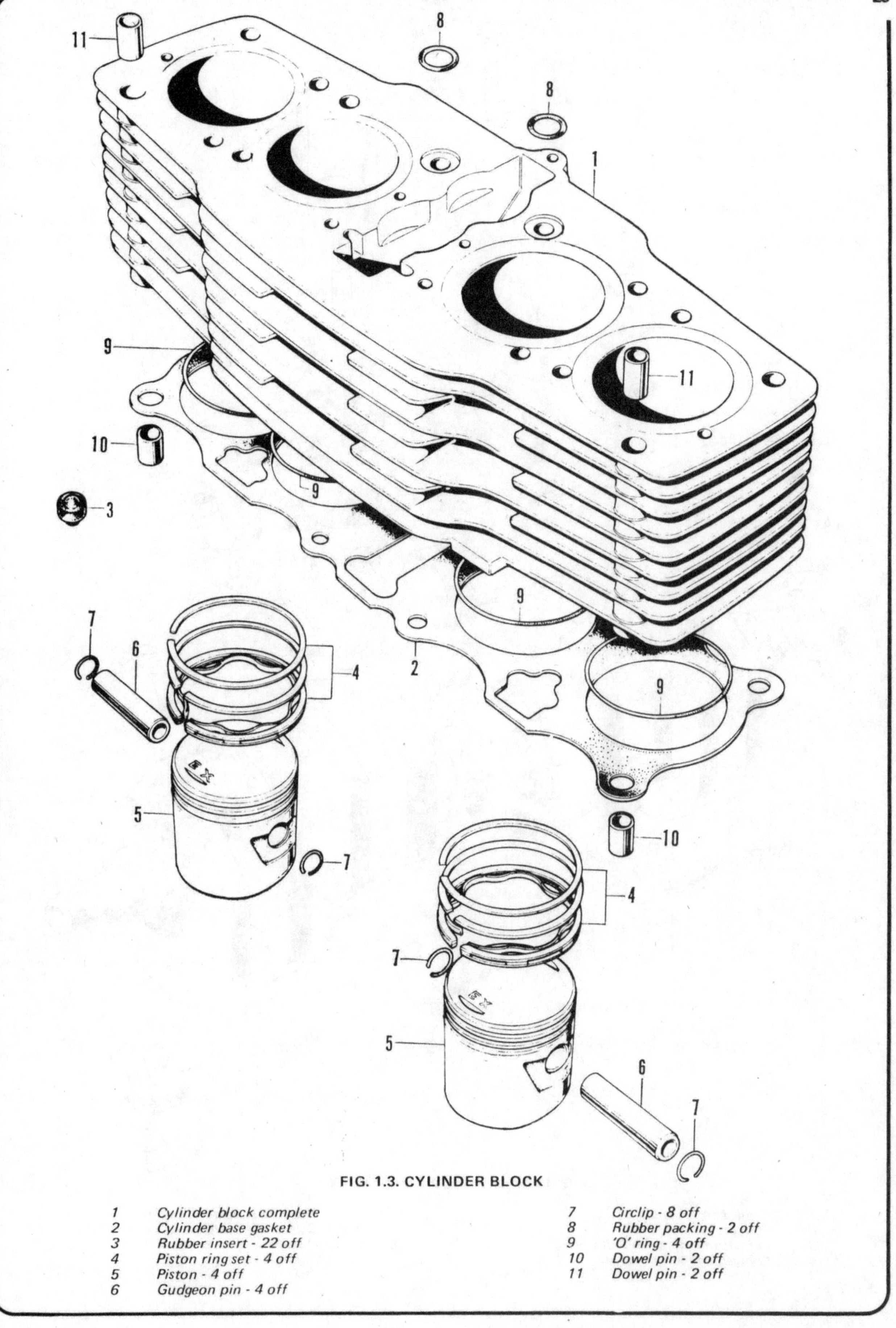

FIG. 1.3. CYLINDER BLOCK

1	*Cylinder block complete*	7	*Circlip - 8 off*
2	*Cylinder base gasket*	8	*Rubber packing - 2 off*
3	*Rubber insert - 22 off*	9	*'O' ring - 4 off*
4	*Piston ring set - 4 off*	10	*Dowel pin - 2 off*
5	*Piston - 4 off*	11	*Dowel pin - 2 off*
6	*Gudgeon pin - 4 off*		

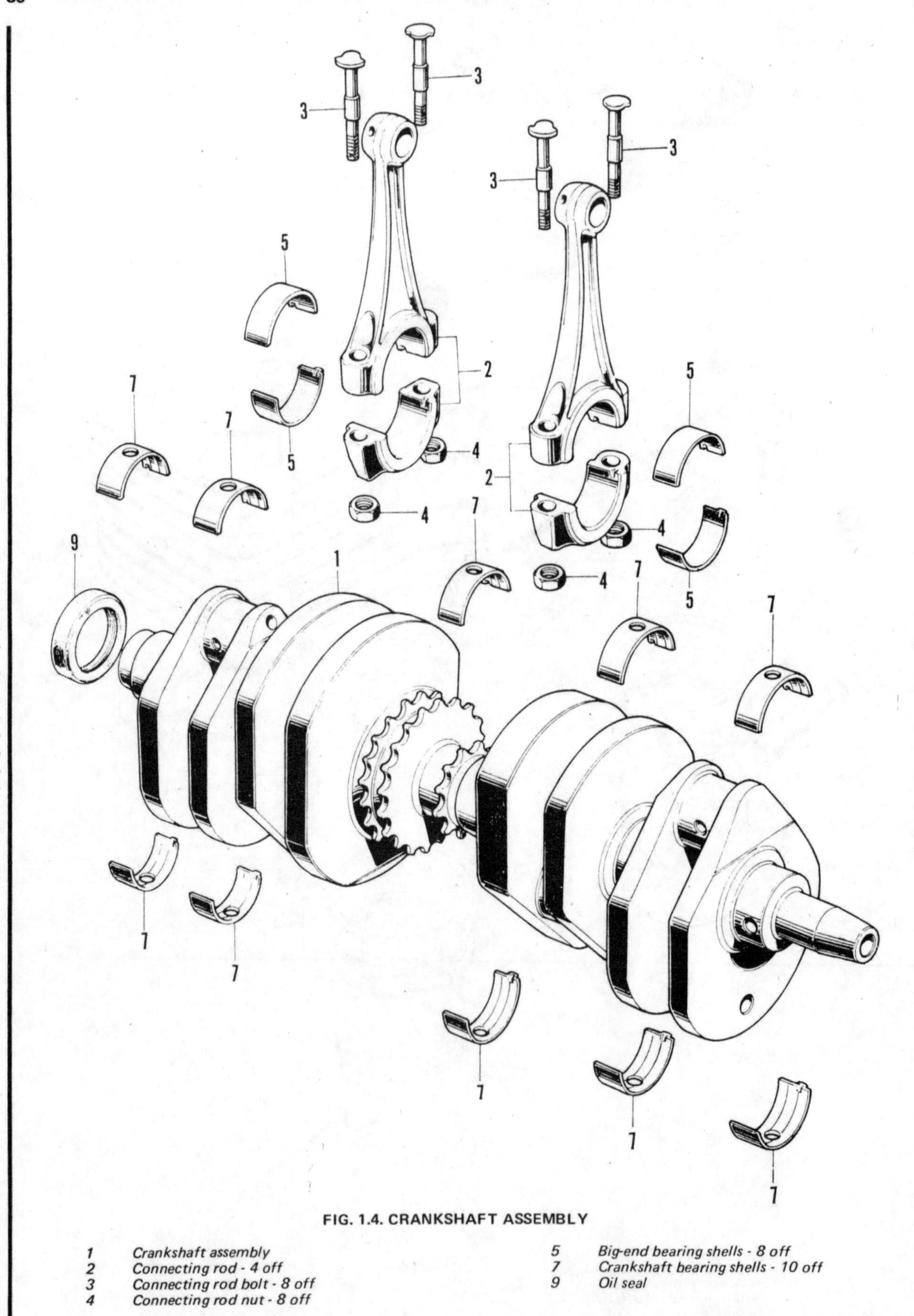

FIG. 1.4. CRANKSHAFT ASSEMBLY

1	*Crankshaft assembly*	*5*	*Big-end bearing shells - 8 off*
2	*Connecting rod - 4 off*	*7*	*Crankshaft bearing shells - 10 off*
3	*Connecting rod bolt - 8 off*	*9*	*Oil seal*
4	*Connecting rod nut - 8 off*		

14.1a. Lift out reduction gear as complete unit

14.2 Engine bolt used to extract selector fork shaft

14.3. Lift out gearbox mainshaft from left

14.3a. Mainshaft sprockets will pull out of endless chain

14.4 Lift out remaining two selector forks

14.4a. Move layshaft to extreme left to detach final drive pinion

14.4b. Take off bottom gear pinion next

14.4c. Pull off top gear pinion next, then ...

14.4d. ... move layshaft to right again and lift from crankcase

14.5. A slotted pin retains the kickstarter shaft

14.5a. Pull the shaft from the splined centre of the pinion, then...

14.5b. ... lift out the pinion assembly, complete with spring

14.7. The primary chain tensioner bolts to the lower crankcase

14.7a. Mark all bearing shells to ensure correct replacement

14.7b. Note 'O' ring around main oil feed dowel

14.7c. Gear selector drum is released by removing neutral contact

16.2. Remove end caps to release connecting rods

16.2a. Tags on bearing shells ensure correct location

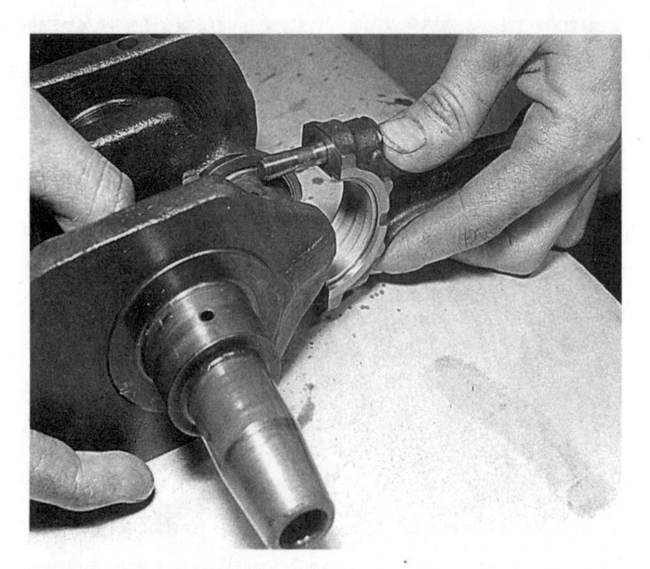

16.4. Wear is evident in the form of score marks or scuffing

18.2. Check small end bearing for blocked oil hole

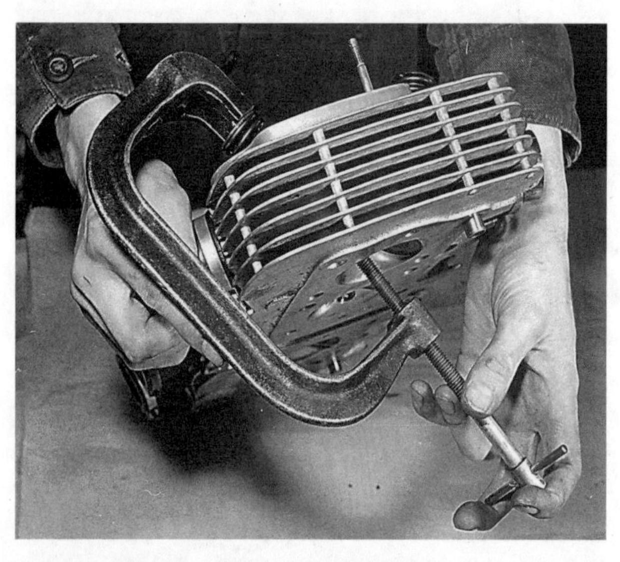

22.2. Use valve spring compressor to detach valves

22.2a. Remove split collets when spring is compressed

22.2b. Each valve has an inner and an outer spring

22.9. Don't forget to replace the spring seatings

22.9a. Fit new oil seals to valve guides

22.9b. Oil valve and guide prior to assembly

22 Cylinder head and valves - dismantling, examination and renovation

1 It is best to remove all carbon deposits from the combustion chambers, before removing the valves for grinding-in. Use a blunt-ended scraper so that the surface of the combustion chambers is not damaged and finish off with metal polish to achieve a smooth, shiny surface.

2 Before the valves can be removed, it is necessary to obtain a valve spring compressor of the correct size. This is necessary to compress each set of valve springs in turn, so that the split collets can be removed from the valve cap and the valve and valve spring assembly released. Keep each set of parts separate; there is no fear of inadvertently interchanging the valves because the heads are of different sizes.

3 Before giving the valves and valve seats further attention, check the clearance between each valve stem and the guide in which it operates. If the measured clearance is greater than 0.003 inch (0.08 mm) for the inlet valve or 0.004 inch (0.1 mm) for the exhaust valve, both the valve and the guide should be renewed. The replacement valve guide should always be one which is oversize so that it can be reamed out to the correct size after it has been inserted in the head. The standard inside diameter for both inlet and exhaust valves is 0.2599 - 0.2603 inch (6.6 - 6.61 mm). To remove the old valve guide, heat the cylinder head and drive it out of position with a double diameter drift of the correct size. Fit the new guide whilst the cylinder head is still warm, using the same drift. A circlip around the body of each guide defines the depth to which the guide can be inserted.

4 Grinding in will be necessary, irrespective of whether new valve guides have been fitted. This action is necessary to remove the indentations in the valve seats caused under normal running conditions by the high temperatures within the combustion chambers. It is also necessary when new valve guides have been fitted in order to re-align the face of each valve with its seating.

5 Valve grinding is a simple task. Commence by smearing a trace of fine valve grinding compound (carborundum paste) on the valve seat and apply a suction tool to the head of the valve. Oil the valve stem and insert the valve in the guide so that the two surfaces to be ground in make contact with one another. With a semi-rotary motion, grind in the valve head to the seat, using a backward and forward action. Lift the valve occasionally so that the grinding compound is distributed evenly. Repeat the application until an unbroken ring of light grey matt finish is obtained on both valve and seat. This denotes the grinding operation is now complete. Before passing to the next valve, make sure that all traces of the valve grinding compound have been removed from both the valve and its seat and that none has entered the valve guide. If this precaution is not observed, rapid wear will take place due to the highly abrasive nature of the carborundum base.

6 When deep pits are encountered, it will be necessary to use a valve refacing machine and a valve seat cutter, set to an angle of 45°. Never resort to excessive grinding because this will only pocket the valves in the head and lead to reduced engine efficiency. If there is any doubt about the condition of a valve, fit a new one.

7 Examine the cc ndition of the valve collets and the groove on the valve stem in which they seat. If there is any sign of damage, new parts should be fitted. Check that the valve spring collar is not cracked. If the collets work loose or the collar splits whilst the engine is running, a valve could drop into the cylinder and cause extensive damage.

8 Check the free length of each of the valve springs. When new, the inner spring has a free length of 1.5 inch (38.1 mm) and the outer spring 1.622 inch (41.2 mm). The springs have reached their serviceable limit when they have compressed to 1.456 inch (37.0 mm) and 1.478 inch (40.0 mm), respective free lengths.

9 Reassemble the valve and valve springs by reversing the dismantling procedure. Fit new oil seals to each valve guide and oil both the valve stem and the valve guide, prior to reassembly. Take special care to ensure the valve guide oil seal is not damaged when the valve is inserted. As a final check after assembly, give the end of each valve stem a light tap with a hammer, to make sure the split collets have located correctly.

10 Check the cylinder head for straightness, especially if it has shown a tendency to leak oil at the cylinder head joint. If the clearance at any point is greater than 0.009 inch (0.25 mm) the cylinder head must be either machined flat or a new head fitted. Most cases of cylinder head warpage can be traced to unequal tensioning of the cylinder head nuts and bolts by tightening them in incorrect sequence.

23 Rocker spindle supports, rocker spindles and rocker arms - examination and renovation

1 Measure the clearance between the rocker arm spindle supports and the rocker arm spindles. If the clearance is greater than 0.005 inch (0.12 mm) renew either the spindle or the spindle support. The choice is best determined by checking the camshaft clearance, as described in the following section, since the spindle support also forms part of the camshaft bearing surface.

2 Check the rocker arms for slackness on the spindle and renew any that show excessive play. Examine the end of each rocker arm where it bears directly on the operating cam and renew any showing signs of scuffing, cracking or any breakthrough into the

specially hardened surface. At the other end of the rocker arm, the internal thread into which the valve adjuster screws must be in good condition, also that of the adjuster itself and the locknut. The tip of the adjuster is hardened; if the tip is cracked, broken or badly scuffed, the complete adjuster must be renewed.

24 Camshaft, camshaft drive chain and tensioner - examination and renovation

1 As mentioned in the previous section, the camshaft is carried in the two rocker arm spindle supports, which are fitted with end caps. When bolted down and tightened to the full torque setting of 6 - 8 lb ft, the clearance between the camshaft and the bearing surface must not exceed 0.008 inch (0.21 mm). Always renew the caps and supports as a complete unit; they are supplied stamped with identical markings.
2 The camshaft should be examined visually for wear, which will most probably be evident on the flanks of each cam and where the cam contour changes sharply. Lift can be checked by measuring the height of the cam from the bottom of the base circle. If the total height of an inlet cam is less than 1.4111 inch (35.86 mm) and for an exhaust cam 1.392 inch (35.36 mm) the camshaft is due for renewal. Evidence of cracking, scuffing or surface breakthrough of any of the cams will also necessitate renewal.
3 Check the camshaft for run-out by supporting it on two V blocks. The run-out should not exceed 0.004 inch (0.10 mm) at any point.
4 It must be emphasised that most cases of rapid wear of the camshaft can be traced to failure to change the engine oil at the recommended period. It is clearly false economy to try and stretch out the period between oil changes; every 2000 miles is a MUST or even more frequently if the machine is used during the winter months or for a whole series of short journeys.
5 The camshaft drive chain can be checked for wear by washing it in petrol, then compressing it endwise so that the free play in both runs is taken up fully. Anchor one end, then pull on the chain so that it stretches as far as possible. If the extension measured exceeds ¼ inch per foot, the chain must be renewed. Do not omit to re-lubricate the chain before reassembly commences. A proprietary lubricant such as Linklyfe or Chainguard is best for this purpose, since the molten lubricant will penetrate the rollers more effectively and, when it has solidified, is less likely to be thrown off by centrifugal action.
6 Wear of the camshaft drive chain guide roller is readily apparent because the centre rib will wear away. The roller is easily detached by withdrawing the rubber plugs from the centre pin, so that the pin can be pushed out.

25 Camshaft drive chain sprockets and tachometer drive - examination and replacement

1 The upper camshaft chain sprocket is bolted to the camshaft and in consequence is easily renewable if the teeth become hooked, worn, chipped or broken. The lower sprocket is integral with the crankshaft and if any of these defects are evident, the complete crankshaft assembly must be renewed. Fortunately, this drastic course of action is rarely necessary, since the parts concerned are fully-enclosed and well-lubricated, working under ideal conditions.
2 If the sprockets are renewed, the chain should be renewed at the same time. It is bad practice to run old and new parts together since the rate of wear will be accelerated.
3 The worm drive to the tachometer is an integral part of the camshaft which meshes with a pinion attached to the cylinder head cover. If the worm is damaged or badly worn, it will be necessary to renew the camshaft complete.

26 Primary drive chains - examination and replacement

1 The two heavy duty primary drive chains should be examined carefully for signs of wear such as damaged rollers or loose side plates. Both chains are of the endless type and under no circumstances should a spring link be included.
2 Wear is best checked before the engine is dismantled or even removed from the frame. The procedure involves draining the oil from the crankcase by removing the drain plug, removing the ten bolts which retain the oil pan in position and then measuring the distance between the primary chain tensioner bracket and the oil pan mounting flange, using a vernier caliper. If the distance between the two points is greater than 2.756 inch (70 mm) the chains must be renewed. Always renew them as a pair, never singly.
3 A check should be made at the same time to ensure both chain sprockets are in good condition. Hooked or badly worn teeth will necessitate renewal, also chipped or broken teeth. The engine sprockets are integral with the crankshaft, in common with the lower camshaft chain sprocket.
4 Whilst the chains are removed from the crankshaft, it is advisable to wash them in a petrol/paraffin mix, permit them to drain off and then immerse them in a molten chain lubricant such as Linklyfe or Chainguard. This will ensure the lubricant penetrates the rollers and when it has solidified, is less likely to be thrown off by centrifugal action.

27 Clutch assembly - examination and renovation

1 The seven clutch friction plates have a bonded lining which gives each plate an overall thickness of 0.1347 - 0.1409 inch (3.42 - 3.58 mm). The serviceable limit of wear is 0.122 inch (3.1 mm). When this point is reached, clutch slip will occur. Measure with a vernier caliper and renew the plates when necessary - they cannot be re-lined. Always renew the plates as a complete set, never singly.
2 Another factor which can promote clutch slip is clutch springs which have taken a permanent set and lost some of their original tension. The free length of each of the four springs is 1.2575 inch (31.94 mm) when new. The serviceable limit is 1.201 inch (3.1 mm) after which replacement of all four springs as a set is essential.
3 Check that none of the clutch plates, either plain or inserted, is buckled. Renewal will be necessary if the plates do not lie flat since it is difficult to straighten them with any measure of success.
4 Examine the clutch assembly for burrs on the edges of the protruding tongues of the inserted plates and/or slots worn in the edges of the outer drum with which they engage. Similar wear can occur between the inner tongues of the plain clutch plates and the slots in the clutch inner drum. Wear of this nature will cause clutch drag and other troubles, since the plates will become trapped and will not free fully when the clutch is withdrawn. A small amount of wear can be treated by dressing with a file; more extensive wear will necessitate renewal of the worn parts.
5 Check the action of the clutch withdrawal mechanism in the clutch cover. It should move freely and return with a positive action. Check the tightness of the ball race in the clutch lifter mechanism. If it is free to revolve in the housing it will be responsible for a loud screech when the drive is taken up. Use Loctite to prevent the bearing from turning in its housing.

28 Gearbox components - examination and renovation

1 Give the gearbox components a close visual inspection for signs of wear or damage such as broken or chipped teeth, worn dogs, damaged or worn splines and bent selectors. Renew any parts found unserviceable because they cannot be reclaimed in a satisfactory manner.
2 Check the condition of the various pawl springs and the kick-starter return spring. If any of these springs fail after the engine has been rebuilt, a further complete stripdown of the engine unit will be necessary.

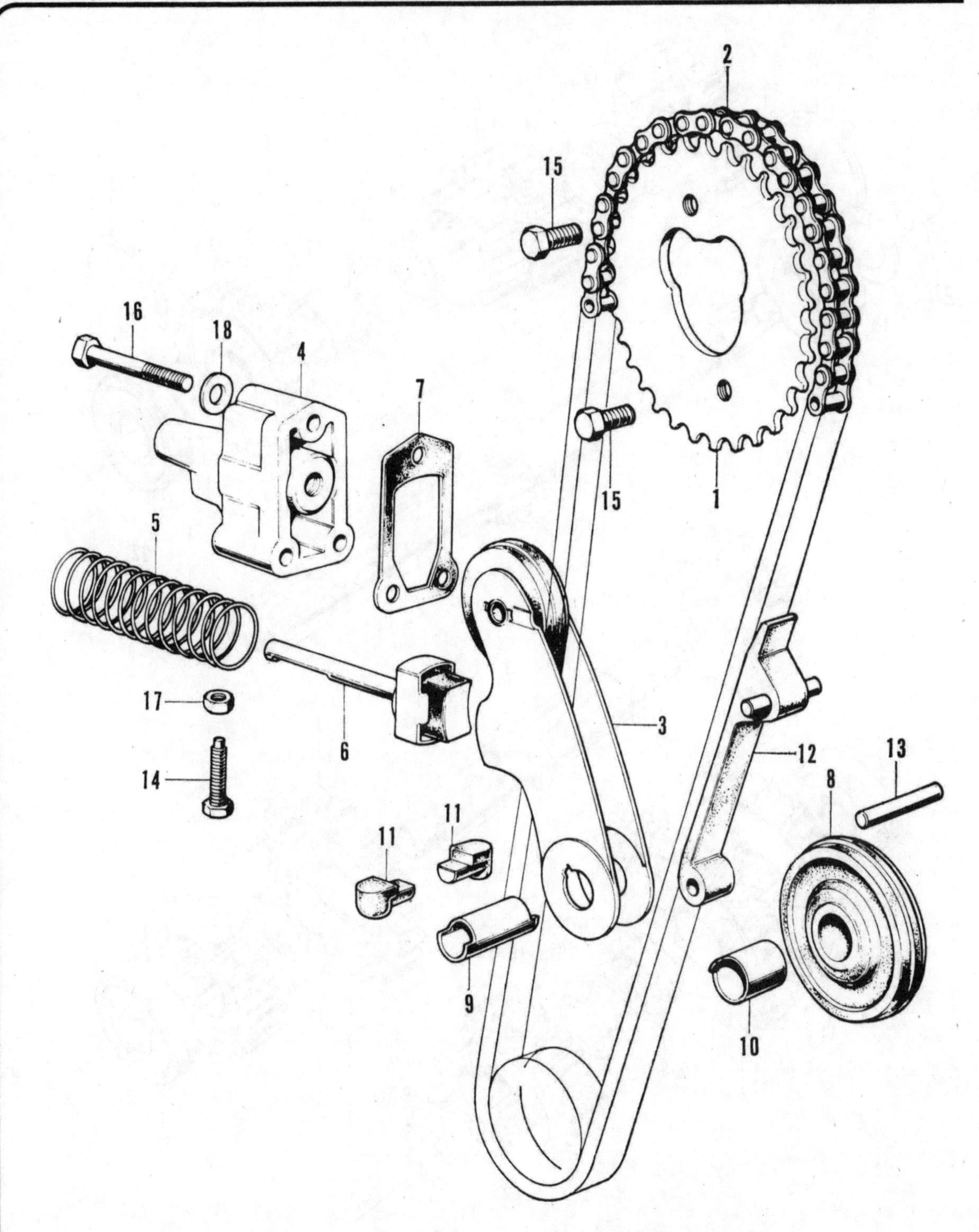

FIG. 1.5. CAMSHAFT CHAIN TENSIONER

1	*Camshaft sprocket*	10	*Camshaft chain roller collar*
2	*Camshaft chain - endless*	11	*Rubber for camshaft chain roller pin - 2 off*
3	*Camshaft chain tensioner*	12	*Camshaft chain guide*
4	*Camshaft chain tensioner holder*	13	*Camshaft chain guide pin*
5	*Tensioner spring*	14	*Tensioner adjusting bolt*
6	*Tensioner push bar*	15	*Sprocket retaining bolt - 2 off*
7	*Tensioner holder gasket*	16	*Bolt - 3 off*
8	*Camshaft chain guide roller*	17	*Nut*
9	*Camshaft chain roller pin*	18	*Plain washer - 3 off*

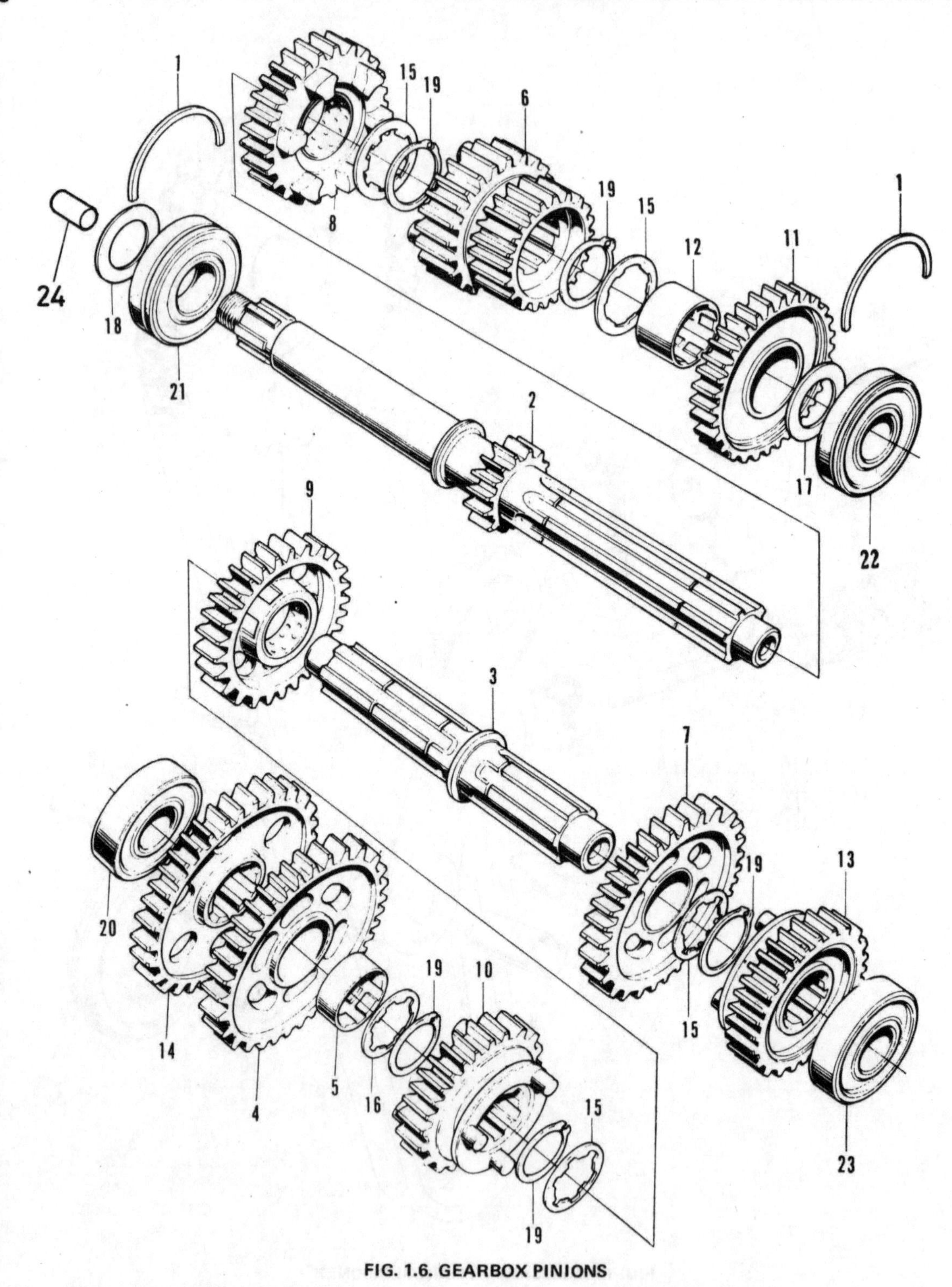

FIG. 1.6. GEARBOX PINIONS

1	*Bearing retaining ring - 2 off*		*13*	*Layshaft top gear (31 teeth)*
2	*Mainshaft (18 teeth)*		*14*	*Final drive gear (48 teeth)*
3	*Layshaft*		*15*	*Thrust washer - 25 mm - 4 off*
4	*Layshaft bottom gear (45 teeth)*		*16*	*Thrust washer - 25 x 33*
5	*Bush*		*17*	*Thrust washer - 20 mm*
6	*Mainshaft 2nd/3rd gear (24/27 teeth)*		*18*	*Thrust washer - 25 mm*
7	*Layshaft 2nd gear (41 teeth)*		*19*	*Circlip - 25 mm - 5 off*
8	*Mainshaft 4th gear (31 teeth)*		*20*	*Radial ball bearing*
9	*Layshaft 3rd gear (36 teeth)*		*21*	*Special ball bearing*
10	*Layshaft 4th gear (34 teeth)*		*22*	*Radial ball bearing*
11	*Mainshaft top gear (33 teeth)*		*23*	*Ball bearing*
12	*Bush*		*24*	*Plug*

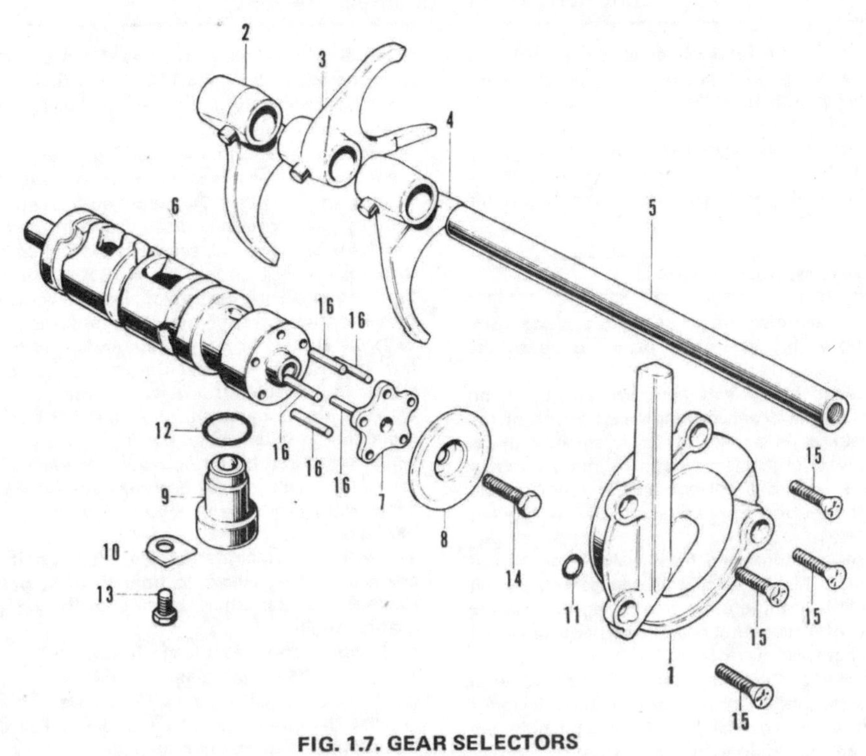

FIG. 1.7. GEAR SELECTORS
(early model shown)

1	*Layshaft bearing retainer*	*9*	*Neutral contact assembly*
2	*Right-hand gear selector fork*	*10*	*Neutral switch stop*
3	*Centre gear selector fork*	*11*	*'O' ring*
4	*Left-hand gear selector fork*	*12*	*'O' ring*
5	*Gear selector fork shaft*	*13*	*Bolt*
6	*Gear selector drum*	*14*	*Bolt*
7	*Gear selector drum stop plate*	*15*	*Countersunk cross-head screw - 4 off*
8	*Gear change side plate*	*16*	*Stop pins - 5 off*

Fig. 1.8. Cylinder head tightening sequence

3 Check the condition of the internal serrations in the kick-starter pinion. If these serrations become worn or the edges rounded, the kickstarter will slip. Renewal of the pinion is the only means of restoring the full action.
4 Check that the tip of the kickstarter pawl is not worn because this too will promote slip during engagement. If the kickstarter pinion is renewed, it is good policy to renew the pawl too.

29 Engine and gearbox reassembly - general

1 Before reassembly is commenced, engine and gearbox components should be thoroughly clean and placed close to the working area.
2 Make sure all traces of old gaskets have been removed and that the mating surfaces are clean and undamaged. One of the best ways to remove old gasket cement, which is needed only on the crankcase and cover joints, is to apply a rag soaked in methylated spirit. This acts as a solvent and will ensure the cement is removed without resort to scraping and the consequent risk of damage.
3 Gather all the necessary tools and have available an oil can filled with clean engine oil. Make sure that all new gaskets and oil seals are available; there is nothing more frustrating than having to stop in the middle of a reassembly sequence because a vital gasket or replacement has been overlooked.
4 Make sure the reassembly area is clean and well lit, with adequate working space. Refer to the torque and clearance settings wherever they are given. Many of the smaller bolts are easily sheared if they are over-tightened. Always use the correct size screwdriver bit for the crosshead screws and NEVER an ordinary screwdriver or punch.

30 Engine and gearbox reassembly - assembling and joining the crankcases

1 Lay the lower crankcase open face uppermost on the work-bench. Build up the layshaft with third, fourth and second gear pinions in position and insert in the crankcase. Push the shaft as far as possible to the left and fit the splined thrust washer, the bush, the bottom gear pinion and the final drive pinion to its right-hand end; ensure both pinions are refitted the correct way round. Insert the shaft into its right-hand bearing and refit the top gear pinion through the left-hand bearing aperture.
2 Insert the gear selector drum and rotate it until the circular depression machined in the wide track at its left-hand end is aligned with the neutral contact aperture; the drum is then in the neutral position. Renew its sealing ring and insert the contact into the crankcase; press it fully into place, fit the stop plate and tighten securely the retaining bolt. Note that the selector forks are marked with the letter L, C or R. Oil the forks and shaft and press the shaft through the bore in the crankcase wall, through all three forks and into the inner crankcase wall. As the shaft is fitted from left to right it passes first through the fork marked L, which is engaged on the layshaft top gear pinion with its guide pin in the drum left-hand track, then through the fork marked C which is fitted with its guide pin in the drum centre track and so that its claw ends project towards the mainshaft, then through the fork marked R which is engaged on the layshaft fourth gear pinion with its guide pin in the drum right-hand track.
3 Insert the kickstarter pinion assembly and shaft, complete with return spring. Hook the end of the return spring over the portion of the crankcase which acts as its retainer, then depress the arm of the kickstarter pawl assembly so that it abuts against the pin acting as its stop. It is best to use a screwdriver for this operation, so that pressure can be maintained whilst the kickstarter shaft is inserted through the centre of the pinion. Note that the shaft will fit the splined centre in one position only, since one spline is specially relieved. Push the shaft home, release the screwdriver and drive the slotted pin back into position so that the shaft is locked in position. A small portion of the pin should protrude, to aid removal on a future occasion. Turn the kickstarter shaft to check that the ratchet assembly functions correctly and that the shaft returns under spring action.
4 Replace the primary chain tensioner in the base of the crankcase. It is retained by a single bolt. Refit the crankshaft bearing shells so that the tagged ends locate correctly, loop both primary drive chains and the camshaft drive chain around their respective sprockets on the crankshaft, then lower the crankshaft into position in the lower crankcase.
5 Reassemble the gearbox mainshaft, complete with the double primary drive sprocket cum shock absorber. Engage this sprocket with the endless primary chain and lower the gearbox mainshaft into position so that the centre selector fork engages with the groove in the second/third gear pinion.
6 Take the upper crankcase, invert it and refit the final drive reduction gear assembly. Do not omit the oil catchment tray which is secured by a single bolt, or the half rings which locate each of the ball journal bearings. The oil seal in front of the left hand bearing must be in good condition if serious oil leaks are to be avoided.
7 Refit the remainder of the bearing half rings to the upper crankcase, using grease to hold them in position, if necessary. Replace the crankshaft bearing shells and give them a liberal coating of oil.
8 Check that all dowels are in position in the lower crankcase and that the large diameter oil path dowel in the front centre is fitted with a new O ring seal. This is most important because the collar channels the main oil feed from the filter to the upper crankcase, under pressure.
9 Coat the jointing faces of both crankcases with gasket cement, then assemble the crankcases together, tapping them to ensure that all bearings and half rings register correctly. It is preferable to have some assistance when the crankcases are assembled together since the connecting rods must be threaded through the upper crankcase whilst it is lowered into position, also the camshaft drive chain.
10 Insert the upper crankcase bolts first, with their respective washers. There are eight 6 mm bolts, two 8 mm bolts and one 10 mm bolt. Tighten them evenly before inverting the crankcase assembly and replacing the bolts in the underside.
11 The underside of the crankcase assembly has ten 6 mm bolts, ten 8 mm bolts, one 8 mm nut and one 10 mm nut, each with their respective washers. The 6 mm bolt nearest the neutral contact carries a clamp for securing the electrical cables.
12 Using a torque wrench tighten the 6 mm bolts to 8 lbf ft (1.1 kgf m) and the 8 mm bolts to 17 – 18 lbf ft (2.3 – 2.4 kgf m). Check that the crankshaft still revolves freely before proceeding further.

31 Engine and gearbox reassembly - assembling the gear change mechanism

1 Before assembling the gear change mechanism, the left hand gearbox layshaft bearing and bearing housing should be replaced as a unit. Engage the end of the layshaft with the bearing centre first, then ease the bearing and bearing housing into position and tighten the four countersunk crosshead screws which retain the assembly in position.
2 Fit the selector shaft retainer first, one end of which passes into the centre of the hollow shaft. It is held by a stud which threads into the crankcase and acts as the pivot for the two stopper arms.
3 Replace the gear change positive stop plate and its return spring. It is retained to the crankcase by a single bolt which passes through a shouldered collar on which the plate pivots and around which the spring is located.
4 Build up the selector stops in the end of the gear selector drum. There are five pins, the shortest of which should be inserted in the deepest hole. Slide the thin stopper plate over the pins, then add the side plate, which is held in position by a central, countersunk screw. It can be replaced in the one position only.

30.1. Build up gearbox layshaft in crankcase

30.1a. Push layshaft to left and add bottom gear pinion, then ...

30.1b. ... add final drive pinion

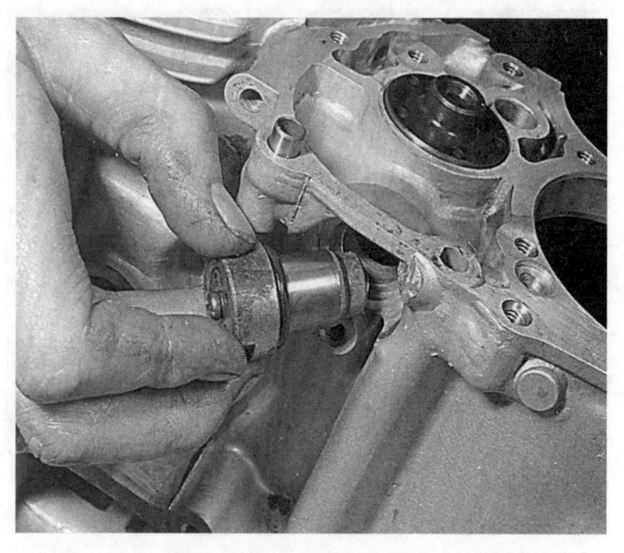

30.2. Insert neutral contact into base of lower crankcase

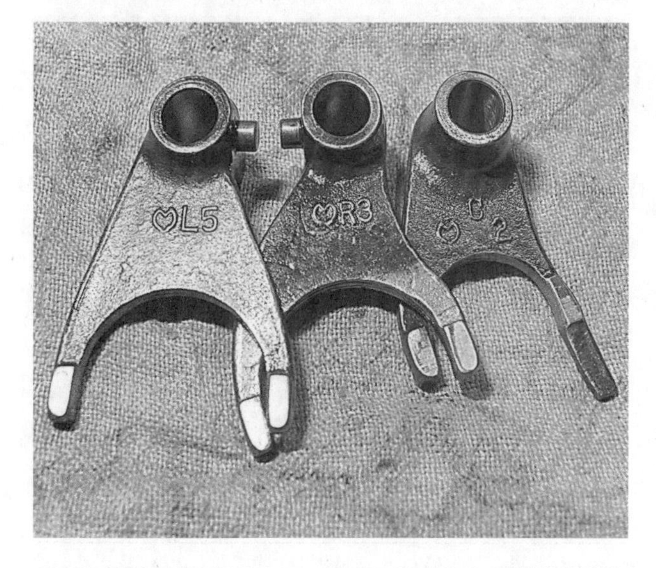

30.2a. Gear selector forks are marked to show their positions

30.2b. Insert selector fork shaft through selector forks

30.3. Insert kickstarter pinion assembly

30.3a. Loop end of return spring over crankcase stop

30.3b. Slotted pin locks shaft in position

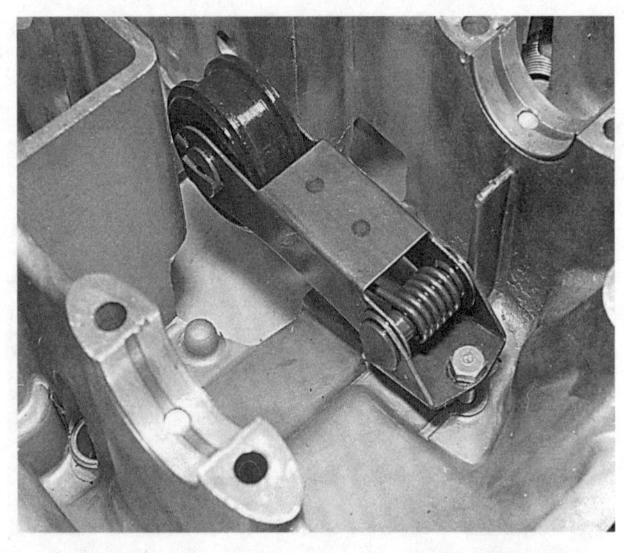

30.4. Replace primary chain tensioner in base of crankcase

30.4a. Refit crankshaft bearing retainers

30.4b. Lower crankshaft into position, complete with chains

30.5. Reassemble mainshaft and engage with primary chain sprockets

30.6. Refit final drive reduction gear in upper crankcase

30.6a. Bolt oil catchment tray into position

30.8. Oil passage dowel must have new 'O' ring

30.9. Keep camshaft chain engaged with crankshaft sprocket during assembly

30.12. Torque wrench must be used to tighten crankcase bolts

31.1. Refit layshaft bearing and holder

31.2. Stud for stopper mechanism secures gear selector shaft

31.3. Replace positive stop plate with spring astride selector drum

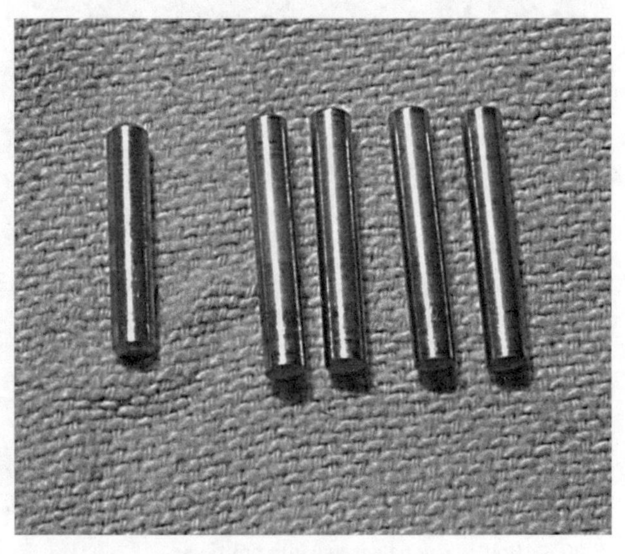

31.4. Selector stops are built up from five pins

31.4a. Build up end plates as shown

31.4b. Tighten, then caulk centre screw

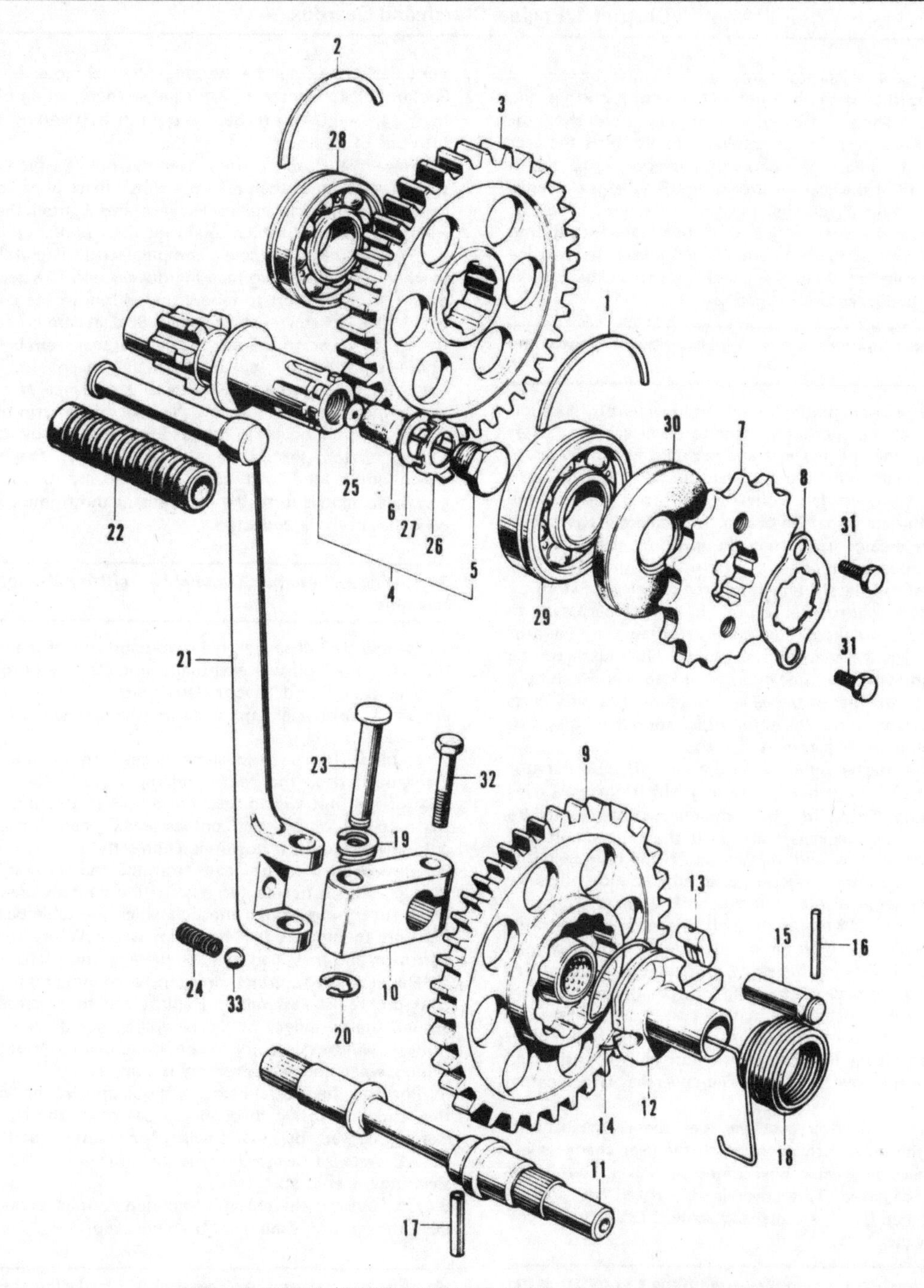

FIG. 1.9. FINAL DRIVE REDUCTION GEAR AND KICKSTARTER

1	*Bearing retaining ring*	*18*	*Kickstarter return spring*
2	*Bearing retaining ring - 3 off*	*19*	*Kickstarter pivot spring*
3	*Final drive gear pinion (56 teeth)*	*20*	*Kickstarter pivot pin clip*
4	*Final drive shaft*	*21*	*Kickstarter arm*
5	*Plug for final drive shaft*	*22*	*Kickstarter arm rubber*
6	*Oil felt*	*23*	*Kickstarter arm pivot pin*
7	*Final drive sprocket (16 tooth *)*	*24*	*Kickstarter pivot stop spring*
8	*Drive sprocket retaining plate*	*25*	*Felt setting washer*
9	*Kickstarter pinion (63 teeth)*	*26*	*Lock washer - 12 mm*
10	*Kickstarter pivot*	*27*	*Sealing washer - 12 mm*
11	*Kickstarter spindle*	*28*	*Special ball bearing*
12	*Kickstarter flange*	*29*	*Ball bearing*
13	*Kickstarter pawl*	*30*	*Oil seal*
14	*Kickstarter ratchet spring*	*31*	*Sprocket retaining bolt - 2 off*
15	*Stop pin*	*32*	*Kickstarter pinch bolt*
16	*Pin*	*33*	*Ball bearing (¼ in. diam.)*
17	*Slotted pin*	*	*Size will vary according to date of manufacture*

5 Replace the gear selector drum stopper and the neutral stopper arm, together with their respective return springs and shouldered collars. The accompanying photographs will show the order in which these parts are assembled. Finally, refit the gear change arm with its integral gear pedal spindle and engage it with the pins in the end of the gear selector drum. Fit the gear change pedal temporarily and check the gear change action. It will be necessary to rotate the gearbox final drive shaft by hand, so that the dogs on the ends of the gear pinions can engage fully as the various gears are selected. When the check has shown that all five gears engage correctly, remove the gear pedal.

32 Engine and gearbox reassembly - replacing the alternator and starter motor drive

1 A free wheel or over-running clutch arrangement is incorporated in the rear of the alternator rotor to disengage the starter motor drive when the engine starts and engine speed increases. It operates centrifugally. When the driven sprocket rotates anticlockwise, the spring-loaded rollers are forced against the crankshaft and the crankshaft is driven. Immediately the crankshaft rotates at a greater speed than the sprocket, the rollers are thrown in the opposite direction, by centrifugal force. The drive is thus disengaged and the starter motor free-wheels to a stop.
2 To inspect the over-running clutch it is first necessary to remove the left hand crankcase cover and withdraw the generator rotor complete with the over-running clutch. The clutch can be dismantled by removing the three countersunk screws which hold the clutch body to the rear of the generator rotor. The only parts likely to need attention are the three roller springs or the bush in the centre of the driven sprocket.
3 Press the large starter pinion into the rear of the alternator rotor, after first checking that the O ring which precedes it, is already in location. Then slide the complete assembly onto the left hand end of the crankshaft and refit the centre bolt and washer. Lock the engine and tighten the rotor bolt securely. Replace the starter motor reduction gear, small pinion outward, followed by the large diameter circular plate which acts as a retainer. The shaft which passes through this assembly is a press fit into the crankcase. If required, the starter motor can be fitted into its enclosure at this stage, but there is no necessity since it does not impede the fitting of the alternator cover assembly. The cover is fitted dry, with a gasket, and is retained by eight crosshead screws. Make sure the rubber grommet carrying the leads from the alternator stator coils is positioned correctly, before the cover is fitted. The cover has two locating dowels.
4 Refit the left hand cover over the gear change mechanism, after renewing the oil seal through which the gear change shaft passes. Care is necessary when positioning the cover, so that the new seal is not damaged. This cover is also fitted dry, with a gasket. It is retained by nine crosshead screws. Do not omit the two locating dowels.

33 Engine and gearbox reassembly - assembling and refitting the clutch

1 Working from the right-hand side of the machine, install the clutch outer drum over the splines of the primary drive sleeve and on later models, secure with the 40 mm circlip. Fit the washer over the mainshaft splines.
2 Build up the clutch inner drum, flanged end downward, by starting with the friction plate which is slightly smaller in diameter than the remainder, then adding the outer clutch ring, tabs facing UPWARD. Build up the remainder of the plates, beginning with one friction plate, then adding one plain plate, until the clutch plate assembly is complete. Finally, add the clutch pressure plate and then fit the inner drum assembly complete, after aligning the tongues of the friction plates to aid engagement with the outer drum. Note that the inner drum will fit the splines in one position only, due to the way in which they are arranged. Replace the spring washer and the lock washer on the mainshaft, then refit the centre sleeve nut, which should be tightened by using the welding rod technique as described in Section 11.4. Tighten to a torque wrench setting of 32.5 - 36.2 lb ft (4.5 - 5.0 kg m). Secure the nut by bending a locking tab into one of its slots.
3 Place the clutch springs over the pillars which extend from the pressure plate then refit the clutch lifter plate complete with centre bearing and collar. Replace and tighten the four bolts, each with washers, which retain the lifter plate in position.
4 Replace the clutch cover, complete with the clutch withdrawal mechanism. It has two locating dowels and is fitted dry, with a gasket. Do not omit to renew the oil seal in the cover, through which the kickstarter shaft passes. Special care is necessary when fitting the cover, to prevent the new oil seal from being damaged. There is also an oil seal in the lifting mechanism itself, which need not be disturbed unless there is evidence of oil leakage. It will be necessary to draw off the clutch lifter arm to gain access.
5 The clutch cover is retained in position by ten crosshead screws, which should be tightened fully. Do NOT fit the chromium plated end cover retained by three countersunk crosshead screws until the engine is in the frame and the clutch cable has been reconnected.

34 Engine and gearbox reassembly - refitting the contact breaker assembly

1 Renew the oil seal at the right hand end of the crankshaft. If this seal leaks, oil will eventually find its way onto the contact breaker points and have a detrimental effect upon the ignition circuit. Take special care to ensure the new seal is fitted without damage.
2 Replace the auto-advance mechanism on the end of the crankshaft, over the centre retaining stud. The dowel in the base of the unit will locate with a hole in the end of the crankshaft, to ensure that the contact breaker cam (integral with the auto-advance unit) is positioned correctly.
3 Thread the electrical leads from the contact breaker assembly through the orifice in the base of the contact breaker housing. Make sure the grommet through which the cable passes is a good tight fit to prevent the ingress of water. Water will give rise to corrosion and may lead to a complete ignition failure.
4 Replace the contact breaker baseplate complete with the contact breaker assembly. Replace the three crosshead screws around the periphery of the baseplate, but do not tighten them fully. It will be necessary to check and possibly reset the ignition timing, when engine reassembly is complete.
5 Position the special hexagon shouldered washer on the end of the stud protruding through the centre of the contact breaker cam, followed by the small plain washer and the 6 mm nut. Beware of overtightening the nut for there is a risk of shearing the stud at its base.
6 Do not refit the chromium plated contact breaker end cover until the ignition timing has been checked.

35 Engine and gearbox reassembly - replacing the pistons and cylinder block

1 Before replacing the pistons, place a clean rag in each crankcase mouth to prevent any displaced component from falling into the crankcase. It is only too easy to drop a circlip whilst it is being inserted into the piston boss, which would necessitate a further stripdown for its retrieval.
2 Fit the pistons in their original order, with the arrow embossed on each crown facing in the forward direction. If the gudgeon pins are a tight fit, warm each piston first, to expand the metal. Do not forget to oil the gudgeon pins and the piston bosses, before fitting.
3 Use new circlips, NEVER the originals. Check that the circlips have located correctly with the groove within each piston boss. A displaced circlip will cause severe engine damage.
4 Fit new O ring seals around the base of the two central holding down studs extending from the rear of the crankcase. Fit a new cylinder base gasket over the crankcase mouths (no

31.5. Replace gear selector drum stopper arm first, followed by ...

31.5a. ... neutral stopper arm

31.5b. End view of completed assembly

31.5c. Refit gear selector arm to engage with stop pins

32.1. Over-running clutch is attached to rear of alternator rotor

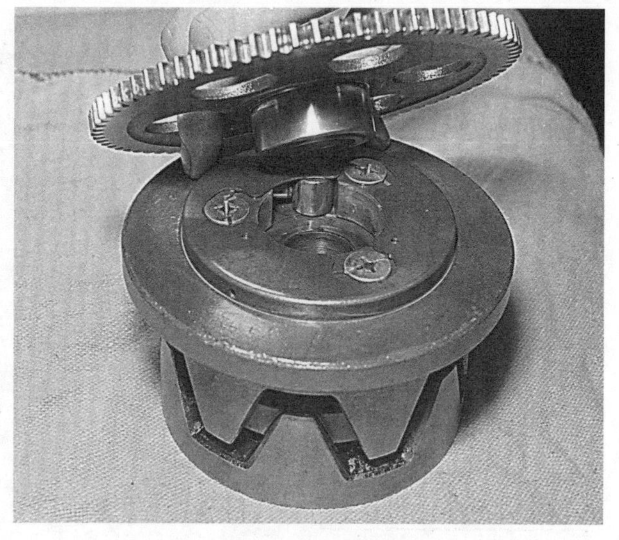
32.3. Press the large starter pinion into the rear of the alternator

32.3a. Fit starter reduction gear with smaller pinion outwards

32.3b. Ease starter motor into position with screwdriver

33.1 Install outer drum on primary drive sleeve splines, secure with circlip on later models, and fit washer

33.2. Build up clutch starting with smaller diameter friction plate and outer ring

33.2a. Continue building up, then fit pressure plate

33.2b. Fit inner drum on splines, then add tab washer and sleeve nut

33.3. Refit the pressure plate after tightening the sleeve nut and replacing the clutch springs

33.4. Replace clutch cover with care, or kickstarter oil seal is damaged

34.2. Auto-advance unit locates with end of crankshaft to ensure correct positioning

34.3. Feed contact breaker lead through orifice first and replace grommet ...

34.3a. ... before positioning base plate

35.2. Arrow on pistons must face forwards

35.3. Warm pistons if gudgeon pins are tight

35.4. New oil seals around centre studs are essential

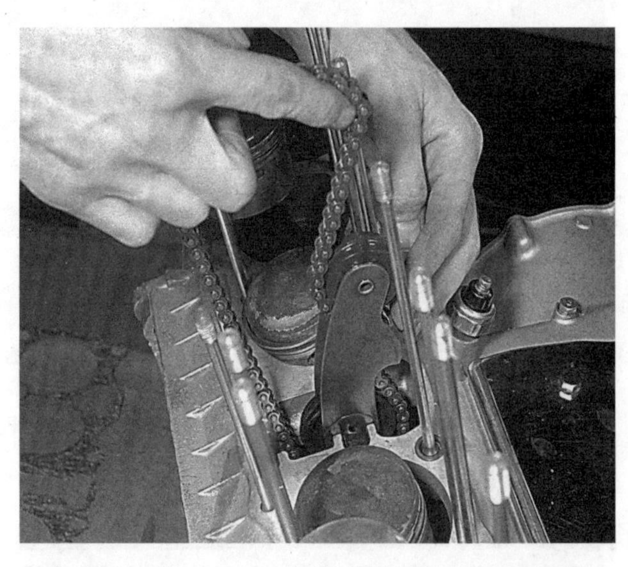

35.4a. Do not forget to thread camshaft chain through tensioner before cylinder block is lowered

36.1. Raise camshaft chain through cylinder head before lowering into position

37.2. Align 'T' mark for cylinders 1 and 4

cement) after checking to ensure that it has been fitted in the correct position and fit new O rings around the cylinder spigot seatings. Do not omit the two dowels which fit around the two outermost holding down studs at the front of the crankcase. Route the camshaft drive chain through the chain tensioner and fit the tensioner in the top of the upper crankcase. Install the mounting rubbers.

5 Position the piston rings so that their end gaps are spaced out of line with each other and fit a pair of piston ring clamps to the two centre pistons. Raise these pistons to the top dead centre position, then smear the bores of the cylinder block with oil and fit the camshaft chain guide and guide pin. Feed the camshaft drive chain upwards through the tunnel in the centre of the block and retain it in this position, then slide the block downward along the holding down studs until the piston ring clamps are displaced and the two centre pistons enter the cylinder bores correctly. Remove the piston ring clamps and fit them to the two outer pistons. Continue to move the cylinder block downward TOGETHER WITH THE TWO CENTRE PISTONS so that the outer pistons will rise from their bottom dead centre position as the crankshaft rotates. Engage these pistons with their cylinder bores, then remove the piston ring clamps and the rag padding from the crankcase mouths before pushing the cylinder block firmly onto the base gasket.

6 It is best to have assistance during this operation, otherwise there is risk of piston ring breakage if the rings do not feed into the bore in a satisfactory manner. It is possible to feed the rings into the bores by hand but this is a somewhat tedious task, it is preferable to use piston ring clamps to simplify the operation.

36 Engine and gearbox reassembly - replacing the cylinder head and tightening down

1 Place the two dowels around the two outer studs projecting through the front of the cylinder block and two new O rings around the two centre studs at the rear of the block. Lay a new cylinder head gasket on top of the block (no cement). Check that the chain is still attached to the sprocket on the crankshaft; feed the camshaft drive chain through the tunnel in the centre of the cylinder head. Lower the cylinder head into position, down the holding down studs, whilst maintaining tension on the camshaft drive chain.

2 Replace the four crosshead screws in the centre of the cylinder head. They are located under the circular sealing rubbers on the outside of each pair of cylinders. There is no screw between each pair of cylinders.

3 Replace the sixteen cylinder head nuts and tighten them down evenly, following the tightening sequence shown in the accompanying illustration. A torque wrench is necessary for the final tightening down, to a setting of 13.7 - 15.2 lb ft (1.9 - 2.1 kg m). Check that the four crosshead screws are tight after this operation, then fit a new circular seal to each of the centre orifices, including the two which do not contain a recessed screw. Do not overlook the two small bolts in the centre (front and rear) of the cylinder head.

37 Engine and gearbox reassembly - replacing the overhead camshaft and timing the valves

1 Replace the two camshaft bearing holders. They slide down the studs which project through the cylinder head.

2 Rotate the engine until pistons Nos 1 and 4 are at top dead centre, then align EXACTLY the T mark for cylinders 1 and 4 with the static timing mark, as viewed through the aperture in the contact breaker baseplate. During this operation, maintain tension on the camshaft drive chain and check that it is still correctly engaged with the crankshaft sprocket.

3 Slide the camshaft chain sprocket on the camshaft, but do not seat it in its final position. Slide the camshaft chain over the right hand end of the camshaft and lower the camshaft into position. It MUST be positioned so that the keyway is in the 12 o'clock position and the scribe line is horizontal, when viewed from the right hand end. FAILURE TO POSITION THE CAMSHAFT IN THIS MANNER WILL RESULT IN INCORRECT VALVE TIMING.

4 Loop the camshaft chain around the camshaft sprocket which is still unattached and then slide the sprocket into position so that it can be bolted to the camshaft flange. The camshaft MUST NOT ROTATE during this operation, so the sprocket may have to be rotated in relation to the chain until the bolt holes align exactly. Tighten the bolts which retain the sprocket, and recheck the alignment of the timing marks. They must register exactly (see accompanying diagram).

38 Engine and gearbox reassembly - replacing the rocker gear and the camshaft chain tensioner

1 Mount the rocker arms on their respective shafts. The shafts and handed pairs of rocker arms are interchangeable, but there should be no problem in replacing them in their original positions if they were marked during the dismantling operation. Replace the bolts which clamp the shafts in position, four in total.

2 Replace the camshaft bearing caps, noting that they are fitted in matched sets. The number stamped on the holder should be identical with that on the cap. Use a torque wrench to tighten the bearing caps to a setting of 6.5 - 9.4 lb ft (0.9 - 1.3 kg m).

3 Replace the camshaft chain tensioner in the rear of the cylinder block. Insert the pushrod and rubber tip first, then the spring, into the tensioner casting and lock the pushrod assembly in the fully compressed position by means of the adjusting screw. Fit the assembled casting to the rear of the cylinder block, using a new gasket at the joint (no cement), then tighten the retaining bolts. Finally, slacken off the locknut and the adjuster bolt so that the pushrod takes up its own adjustment of the chain. Retighten the adjuster locknut; the chain tension is now correct.

39 Engine and gearbox reassembly - checking the valve clearances

1 The valve clearances must always be checked with the engine COLD. Commence by turning the engine clockwise (normal direction of rotation) and watch the inlet valve tappet of No 1 cylinder (extreme left hand cylinder). Continue turning the engine until the inlet valve has opened and then closed, then align the timing marks viewed through the aperture in the contact breaker baseplate so that the T mark for cylinders 1 and 4 is exactly in line with the static timing mark. No 1 piston is now exactly at top dead centre and both valves are fully closed. The valve clearances of No 1 cylinder can now be checked by means of a feeler gauge and the inlet valve of No 3 cylinder and the exhaust valve of No 2 cylinder - which are also closed in this engine position.

2 The standard valve clearance is 0.002 inch (0.05 mm) for the inlet valves and 0.003 inch (0.08 mm) for the exhaust valves. Adjustment is made by slackening the locknut at the end of the rocker arm and by turning the adjusting screw until the feeler gauge is a good sliding fit. Hold the adjusting screw whilst the locknut is retightened, then recheck the valve clearance.

3 Rotate the engine one complete turn (360°) until the piston in No 4 cylinder is exactly at top dead centre, by following an identical procedure. Both valves on No 4 cylinder can be checked, also the remaining valves on the other cylinders - inlet valve on No 2 and the exhaust valve on No 3. Recheck the settings after adjustment and tightening the locknuts.

4 It is most important that the valve clearances are checked when the valves are fully closed. If adjustments are made when a valve is still slightly on lift, an excessive clearance will result, making the valve gear noisy and causing loss of performance. NEVER set the clearances less than the recommended values. If the valves do not seat correctly when the engine is hot, both valves and seatings will burn away, necessitating an expensive top overhaul of the engine. It is never practicable to quieten a

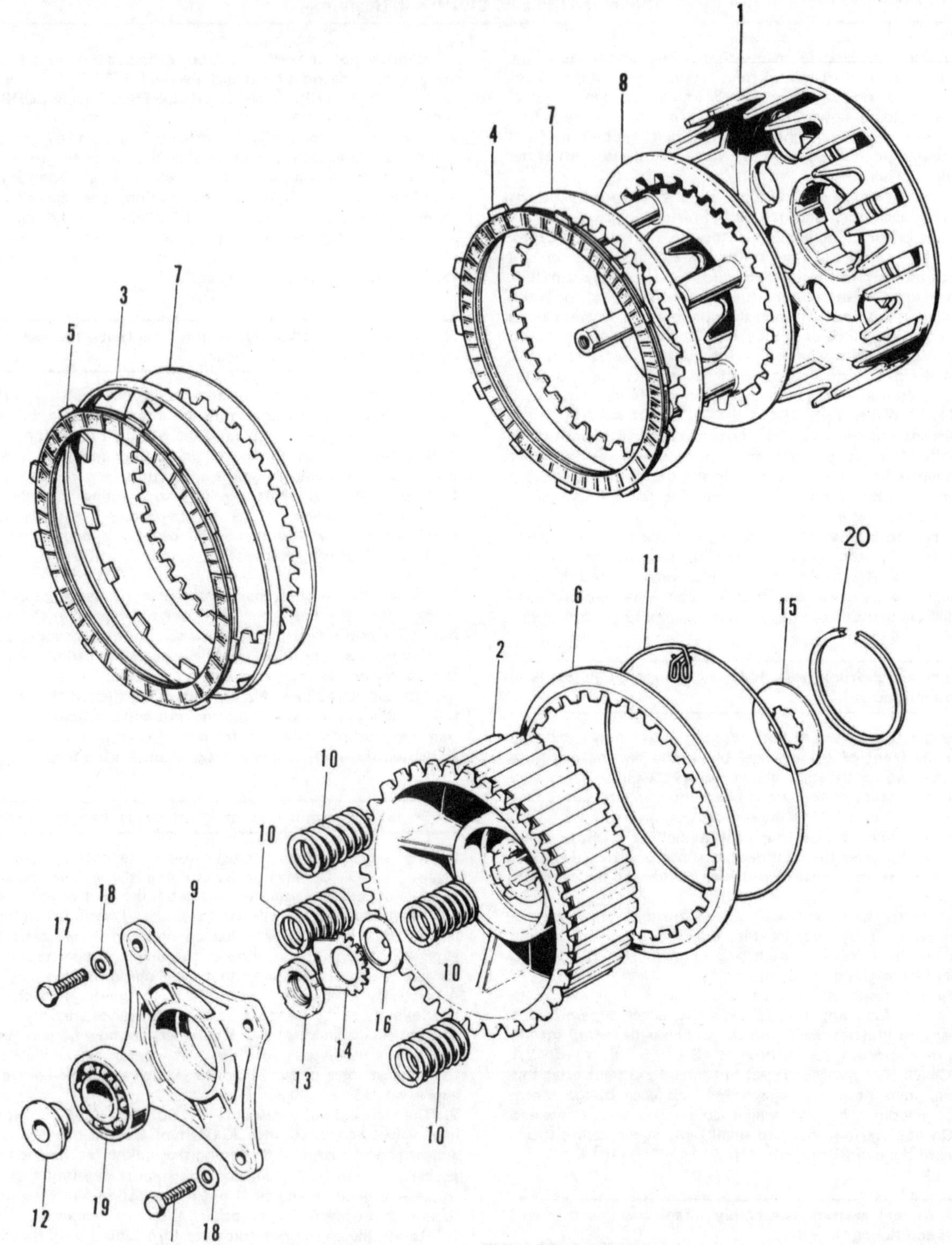

FIG. 1.10. CLUTCH

1 Clutch outer drum
2 Clutch inner drum
3 Clutch outer ring (early models)
4 Friction plates
5 Inner friction plate
6 Inner plain plate (early models)
7 Plain plates
8 Pressure plate
9 Clutch lifter plate
10 Clutch spring – 4 off
11 Clutch plate stopper ring (early models)
12 Clutch lifter insert
13 Sleeve nut
14 Lock washer
15 Splined washer
16 Spring washer
17 Bolt – 4 off
18 Washer – 4 off
19 Bearing
20 40 mm circlip (later models)

37.3. Right-hand end of camshaft must be aligned as shown

37.4. Loop chain around sprocket so that bolt holes align WITHOUT moving camshaft

38.1. Insert rocker shafts with slotted end facing outwards

39.1. Always check valve clearances with engine in recommended position

40.1. Two cylinder head cover screws are within breather cover

41.2. Note rag protecting frame enamel during engine unit lift-in

41.3. Exhaust stubs are retained by two cross head screws

41.8. Projection on filter body engages with slot in crankcase casting

41.9. Fit new 'O' ring seals at oil pipe union joints

41.10. Replace the carburettors complete with mounting plate

41.10a. Don't forget to tighten clamps

41.10b. Air cleaner box bolts to two frame lugs

noisy engine in this manner. If the valve clearances are checked when the engine is in the positions indicated, this will ensure the valves concerned are fully closed.

40 Engine and gearbox reassembly - refitting the cylinder head cover

1 Place a new gasket on the cylinder head flange (no cement) and replace the cylinder head cover, which is retained by fourteen crosshead screws. Two of these are within the area enclosed by the breather cover. Tighten them down evenly.
2 Before the cylinder head cover is replaced and tightened down, it is advisable to lubricate the camshaft and the rocker gear with oil from an oil can. This will prevent any of the valve gear components from running dry during the initial start up, since it will take a few minutes for oil pressure to build up and the oil to circulate to some of the areas more remote from the oil pump.
3 Replace the breather cover in the top of the cylinder head cover. A new gasket should be fitted at the joint (no cement) and the breather cover tightened down by tightening the three crosshead screws which retain it in position.

41 Replacing the engine and gearbox unit in the frame

1 The engine and gearbox unit should be lifted into the frame from the right hand side, with the rear end lifted slightly upward. This is a task which requires two persons and it is important that the machine is standing firmly on level ground, resting on the centre stand.
2 Since the engine unit is a tight fit in the frame, it is advisable to wrap some rag around the two front down tubes of the frame to prevent them from being scratched as the engine is moved into its correct location. It will also help to prevent the risk of damage to any of the cylinder head fins.
3 Before the engine is located in its final position and the engine bolts inserted, it is advisable to refit the three exhaust stubs to the exhaust ports. Each is retained by two countersunk crosshead screws, which must be tightened fully. It is possible to gain better access with a screwdriver at this stage, particularly in the case of the stub attached to No 1 cylinder. Do not omit to fit a new copper sealing washer between each stub and the exhaust port.
4 Support the engine unit from the underside and locate the long engine bolt which passes through the lugs welded to the lower frame tubes, on the underside of the engine. Follow up with the long bolt through the upper rear portion of the crankcase casting the rod (which carries the footrests) and then the remaining short bolts, some of which pass through the small detachable engine plates fitted to the front and rear of the engine unit on the right hand side only. When all the bolts are positioned correctly, tighten them fully.
5 Re-attach the breather pipe to the rear of the cylinder head and retain it in position with the wire clip, also the breather pipe attached to the rear of the upper crankcase. Thread the starter motor cable through the frame and re-attach it to the starter solenoid switch in the electrical compartment. Do not omit the insulating cover over the terminal when the connection has been made and tightened.
6 Rejoin the electrical connections at the junction box in front of the battery and the lead to the stop lamp switch. Do not omit to remake the connections to the neutral contact on the underside of the engine and to the oil pressure switch, on top of the crankcase. The contact breaker leads connect with the snap connector near the oil tank.
7 Connect the clutch cable, ensuring that it is correctly routed, then slacken their locknuts and screw in fully both cable adjusters to gain the maximum cable free play. Slacken its locknut and unscrew (clockwise) by one full turn the clutch release mechanism adjuster screw to ensure there is no pressure on it. Screw in (anti-clockwise) the adjuster screw until light resistance

41.11. Place finned clip on exhaust stubs first, before fitting pipe

41.13. Sprocket is retained by locking plate and two bolts

41.13a. Sprocket cover cannot be fitted with chain in position

is encountered, then set the necessary clearance by unscrewing it (clockwise) through ¼ - ½ turn. Hold the screw and tighten securely the locknut. Refit the chrome-plated inspection cover and use the cable lower adjuster to set the specified cable free play of 10 - 25 mm (0.4 - 1.0 in) measured in terms of free movement at the handlebar lever ball end. Reserve the handlebar adjuster for fine roadside adjustments. When the clutch is adjusted correctly and operates with no signs of slip or drag, tighten the adjuster locknuts and replace the rubber sleeve over the cable lower end. Refit the kickstart lever and gearchange pedal, tightening securely their pinch bolts.

8 Refit the oil filter unit to the front of the crankcase, using a new element. Make sure the rubber O ring seal fitted to the rim of the cover is in good condition before replacing the centre bolt and tightening it fully. If in doubt, renew the O ring because this joint may otherwise leak quite badly. Note that a projection on the filter body engages with a slot cast on the crankcase.

9 Replace the oil tank, which is held in position by three bolts with captive nuts. Rejoin the oil pipes with their connections at the base of the lower crankcase, fitting new O ring seals to the jointing flanges. When the flanges have been tightened, refill the tank with the correct grade of oil. It will hold 6.2 Imp pints (3.5 litres).

10 Replace the carburettors, together with their mounting plate. When both inlet and outlet hoses have been connected, tighten the retaining clamps. Early models having separate throttle cables should have the individual slide and needle assemblies inserted at this stage, so that the carburettor tops can be tightened down. Later models will have the tops already in position, so that only the linkage need be reconnected. Refit the air cleaner box complete with filter element. It is retained by two bolts, through frame lugs.

11 Refit the exhaust pipes and silencers, making sure that the short balance pipes joining each pair of silencers have connected correctly. It is necessary to place the finned clips on the exhaust stubs first, then slide them onto the pipes. Give the pillion footrest bolts a final tightening after the finned exhaust pipe clamps have been tightened.

12 Replace the petrol tank and reconnect the fuel lines to the carburettors. Make sure that the carburettor drain pipes are positioned so that they will not flood any of the engine components, especially any part of the electrical equipment.

13 Replace the final drive gearbox sprocket if it has been detached previously, followed by the locking plate and two bolts. Replace the cover around the gearbox sprocket, held by two bolts. If the final drive chain is of the endless type it can be looped around the rear wheel sprocket and fitted in conjunction with the gearbox sprocket. It may be necessary to move the rear wheel forward temporarily to gain enough slack for the chain and sprocket to be fitted to the final drive shaft in conjunction with one another. Earlier models have a chain which incorporates a spring link, which is much easier to fit. Position the ends of the chain in the teeth of the rear wheel sprocket to make the insertion of the spring link easier. The closed end of the spring clip MUST face the direction of travel of the chain and must engage correctly with the grooves in the chain link.

14 Reconnect the positive terminal of the battery and give the electrical system a quick visual check to ensure it is functioning correctly. Replace the side cover over the electrical compartment and the side cover over the oil tank. They are a push-on fit. Refit the cover over the final drive assembly, which is retained by two crosshead screws and two dowels.

15 Before starting the engine, the ignition timing must be checked and if necessary reset, according to the procedure given in Chapter 3, Section 7. When the setting is correct for both pairs of cylinders, tighten the screws around the periphery of the baseplate, then replace the contact breaker end cover, which is retained by two countersunk crosshead screws. Make sure the cover has a gasket to seal the joint with the contact breaker housing. If water enters, an ignition failure will occur.

42 Starting and running the rebuilt engine

1 Open the petrol tap, close the carburettor chokes and start the engine, using either the kickstarter or the electric starter. Raise the chokes as soon as the engine will run evenly and keep it running at a low speed for a few minutes to allow oil pressure to build up and the oil to circulate. If the red oil pressure indicator lamp is not extinguished, stop the engine immediately and investigate the lack of oil pressure.

2 The engine may tend to smoke through the exhausts initially, due to the amount of oil used when assembling the various components. The excess of oil should gradually burn away as the engine settles down.

3 Check for the return feed of oil to the oil tank by removing the filler cap and observing the return flow. Initially the flow should be free from air bubbles, but eventually air bubbles will appear as the scavenge pump clears the crankcase of excess oil. The scavenge pump has a greater capacity than the delivery pump, hence the presence of air. If there is no flow, check the main feed pipe for an air lock.

4 Check the exterior of the machine for oil leaks or blowing gaskets. Make sure that each gear engages correctly and that all the controls function effectively, particularly the brakes. This is an essential last check before taking the machine on the road.

43 Taking the rebuilt machine on the road

1 Any rebuilt machine will need time to settle down, even if parts have been replaced in their original order. For this reason it is highly advisable to treat the machine gently for the first few miles to ensure oil has circulated throughout the lubrication system and that any new parts fitted have begun to bed down.

2 Even greater care is necessary if the engine has been rebored or if a new crankshaft has been fitted. In the case of a rebore, the engine will have to be run-in again, as if the machine were new. This means greater use of the gearbox and a restraining hand on the throttle until at least 500 miles have been covered. There is no point in keeping to any set speed limit; the main requirement is to keep a light loading on the engine and to gradually work up performance until the 500 mile mark is reached. These recommendations can be lessened to an extent when only a new crankshaft is fitted. Experience is the best guide since it is easy to tell when an engine is running freely.

3 If at any time a lubrication failure is suspected, stop the engine immediately, and investigate the cause. If an engine is run without oil, even for a short period, irreparable engine damage is inevitable.

4 When the engine has cooled down completely after the initial run, recheck the various settings, especially the valve clearances. During the run most of the engine components will have settled into their normal working locations.

44 Fault diagnosis - engine

Symptom	Reason/s	Remedy
Engine will not start	Defective spark plugs	Remove the plugs and lay on cylinder heads. Check whether spark occurs when ignition is switched on and engine rotated.
	Dirty or closed contact breaker points	Check condition of points and whether gap is correct.
	Faulty or disconnected condenser	Check whether points arc when separated. Replace condenser if evidence of arcing.
Engine runs unevenly	Ignition and/or fuel system fault	Check each system independently, as though engine will not start.
	Blowing cylinder head gasket	Leak should be evident from oil leakage where gas escapes.
	Incorrect ignition timing	Check accuracy and if necessary reset.
Lack of power	Fault in fuel system or incorrect ignition timing	See above.
Heavy oil consumption	Cylinder barrels in need of rebore	Check for bore wear, rebore and fit oversize pistons if required.
	Damaged oil seals	Check engine for oil leaks.
Excessive mechanical noise	Worn cylinder barrels (piston slap)	Rebore and fit oversize pistons.
	Worn camshaft drive chain (rattle)	Adjust tensioner or replace chain.
	Worn big end bearings (knock)	Fit replacement crankshaft assembly.
	Worn main bearings (rumble)	Fit new journal bearings and seals. Replace crankshaft assembly if centre bearings are worn.
Engine overheats and fades	Lubrication failure	Stop engine and check whether internal parts are receiving oil. Check oil level in crankcase.

45 Fault diagnosis - gearbox

Symptom	Reason/s	Remedy
Difficulty in engaging gears	Selector forks bent	Renew.
	Gear clusters not assembled correctly	Check gear cluster arrangement and position of thrust washers.
Machine jumps out of gear	Worn dogs on ends of gear pinions	Renew worn pinions.
	Stopper arms not seating correctly	Remove right hand crankcase cover and check stopper arm action.
Gear change lever does not return to original position	Broken return spring	Renew spring.
Kickstarter does not return when engine is turned over or started	Broken or poorly tensioned return spring	Renew spring or re-tension.
Kickstarter slips	Ratchet assembly worn	Part crankcase and renew all worn parts.

46 Fault diagnosis - clutch

Symptom	Reason/s	Remedy
Engine speed increases as shown by tachometer but machine does not respond	Clutch slip	Check clutch adjustment for free play at handlebar lever. Check thickness of inserted plates.
Difficulty in engaging gears. Gear changes jerky and machine creeps forward when clutch is withdrawn. Difficulty in selecting neutral	Clutch drag	Check clutch adjustment for too much free play. Check clutch drums for indentations in slots and clutch plates for burrs on tongues. Dress with file if damage not too great.
Clutch operation stiff	Damaged, trapped or frayed control cable	Check cable and renew if necessary. Make sure cable is lubricated and has no sharp bends.

Chapter 2 Fuel System and Lubrication

Contents

Specifications

Fuel tank capacity	4.0 Imp gallons/5.0 US gallons/18.0 litres
Carburettors	
Make/type	Keihin/piston valve
Choke size	28 mm
Main jet	120
Pilot jet	100
Air bleed	0.035 in x 4 Carburettors 1 and 2 0.024 in x 2 Carburettors 3 and 4
Needle jet	0.012 x 0.15 in (7.6 x 3.8 mm)
Jet needle	0.098 in (2.485 mm)
Cutaway (throttle valve)	2.5
Pilot jet setting	1 - 1$^1/_8$ turns out
Slow running jet	40
Oil tank capacity	5.2 Imp pints/7.4 US pints/3.0 litres
Oil pressure	56.9 psi @ 4000 rpm, oil temperature 80°C
Clearance, oil pump outer rotor to pump body	0.0138 in max (0.35 mm)
Clearance between rotors	0.0138 in max (0.35 mm)
Clearance, rotors to pump body (endwise)	0.0047 in max (0.12 mm)

1 General description

The fuel system comprises a petrol tank from which petrol is fed by gravity to the float chamber of each of the four carburettors. A single petrol tap with a detachable gauze filter is located beneath the petrol tank, on the right hand side. It contains provision for a reserve quantity of petrol, when the main supply is exhausted.

For cold starting, a hand-operated choke lever attached to the far left hand carburettor is linked to the three other carburettors so that the mixture can be enrichened temporarily. When the engine has started, the choke can be opened gradually as the engine warms up, until full air is accepted under normal running conditions. On early models, each carburettor is operated by its own separate control cable. Later models employ a push-pull cable arrangement which actuates a mechanical linkage attached to each carburettor.

Lubrication is effected by the dry sump principle, in which oil from the separate side-mounted oil tank is delivered by gravity to the mechanical oil pump located within the sump plate, at the base of the crankcase. The pump is of the trochoid type and is driven from the gearbox mainshaft via the kick-starter pinion. Oil is supplied under pressure via a pressure release valve and a full flow oil filter with a replaceable element to the crankshaft and to the overhead camshaft and rocker gear. A secondary flow passes to the gearbox via the gearbox main bearings. All surplus oil drains to the sump and is returned to the oil tank by the scavenge section of the oil pump. The pump itself is protected by a gauze strainer in the base, and a ball valve is incorporated in the system to shut off the feed from the oil tank when the engine stops.

2 Petrol tank - removal and replacement

1 The petrol tank is retained by two guide channels which locate with a circular rubber block on each side of the steering head and a rubber strap which fits over a lip projecting from the rear of the petrol tank.

2 To remove the petrol tank, turn off the petrol tap and detach both petrol pipes, after first releasing the wire retaining clips. Raise the seat, pull the rubber strap away from the lip at

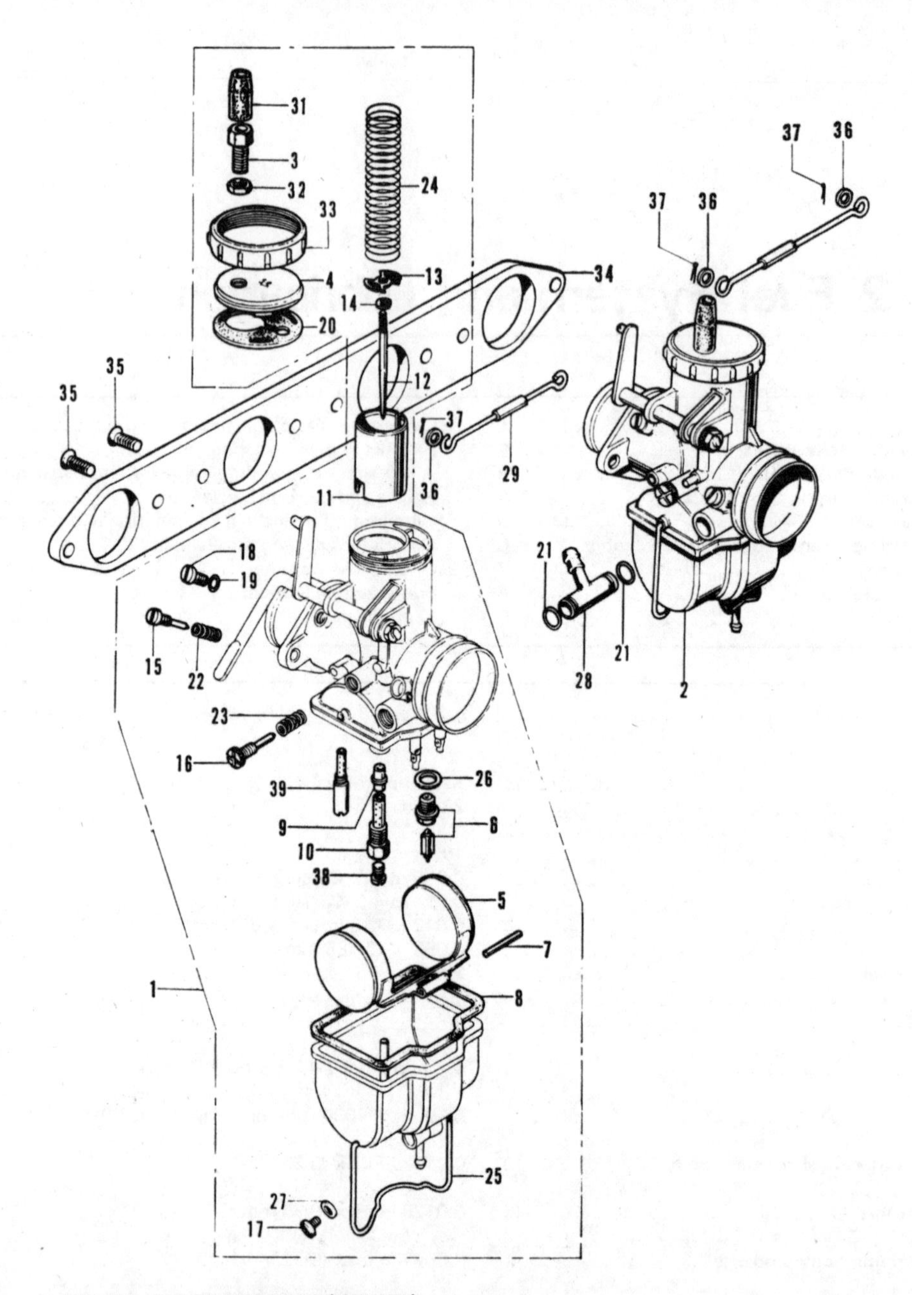

FIG. 2.1. CARBURETTORS (KEIHIN)

1 Carburettor complete - No. 1 cylinder
2 Carburettor complete - No. 2 cylinder *
3 Cable adjuster - 4 off
4 Carburettor top - 4 off
5 Float assembly - 4 off
6 Float needle and seat - 4 off
7 Hinge pin for float - 4 off
8 Float chamber gasket - 4 off
9 Needle jet - 4 off
10 Needle jet holder - 4 off
11 Throttle valve - 4 off
12 Needle - 4 off
13 Needle clip plate - 4 off
14 Needle clip - 4 off
15 Air screw - 4 off
16 Throttle stop screw - 4 off
17 Float chamber drain screw - 4 off
18 Vacuum gauge blanking screw - 4 off
19 Plain washer - 4 off
20 Carburettor top gasket - 4 off
21 'O' ring - 4 off
22 Air screw spring - 4 off
23 Throttle stop screw spring - 4 off
24 Throttle spring - 4 off
25 Float chamber retaining clip - 4 off
26 Plain washer - 4 off
27 Plain washer - 4 off
28 Fuel pipe 'T' junction - 2 off
29 Choke rod assembly A
30 Choke rod assembly B - 3 off
31 Rubber cap - 4 off
32 Nut - 4 off
33 Locking ring - 4 off
34 Mounting plate
35 Countersunk cross head screws - 8 off
36 Plain washer - 6 off
37 Split pin - 2 off
38 Main jet - 4 off
39 Slow running jet - 4 off
* Carburettors for cylinders 3 & 4 are identical but right-handed

5.4. Withdraw split pin to release choke connection

6.1. Withdraw hinge pin to release twin float assembly

the rear of the tank, and lift the tank away.

3 When replacing the tank, by reversing the above procedure, make sure the tank seats correctly and does not trap any of the control cables, especially those of the carburettors.

3 Petrol tap - removal and replacement

1 If only the petrol tap filter requires attention, there is no necessity to remove the tap or to drain the petrol tank. The filter bowl, which has a hexagon head to aid removal, is threaded into the base of the petrol tap and can be unscrewed after the tap has been turned to the 'off' position. The circular filter cause will also be released and can be washed with petrol to remove any sediment. Before replacing, the filter bowl should be cleaned thoroughly.

2 It is seldom necessary to remove the lever which operates the petrol tap, although occasions may occur when a leakage develops at the joint. Although the tank must be drained before the lever assembly can be removed, there is no need to disturb the body of the tap.

3 To dismantle the lever assembly, remove the two crosshead screws passing through the plate on which the operating positions are inscribed. The plate can then be lifted away, followed by a spring, the lever itself and the seal behind the lever. The seal will have to be renewed if leakage has occurred. Reassemble the tap in the reverse order. Gasket cement or any other sealing medium is NOT necessary to secure a petrol tight seal.

4 If the tap body has to be removed, it is held to the underside of the petrol tank by two crosshead screws with washers. Note that there is an O ring seal between the petrol tap body and the petrol tank, which must be renewed if it is damaged or if petrol leakage has occurred.

4 Petrol feed pipes - examination

1 Synthetic rubber feed pipes are used to convey the flow of petrol from the petrol tap to the float chamber of each of the four carburettors. Each pipe is retained by a wire clip, which must hold the pipe firmly in position. Check periodically to ensure the pipes have not begun to split or crack and that the wire clips have not worn through the surface.

2 Do NOT replace a broken pipe with one of natural rubber, even temporarily. Petrol causes natural rubber to swell very rapidly and disintegrate, with the result that minute particles of rubber would easily pass into the carburettors and cause blockages of the internal passageways. Plastic pipe of the correct bore size can be used as a temporary substitute but it should be replaced with the correct type of tubing as soon as possible since it will not have the same degree of flexibility.

5 Carburettors - removal

1 On early models having a separate throttle cable for each carburettor, unscrew each carburettor top and lift out the throttle valve, return spring and needle as a complete assembly. Tape each assembly to an adjacent frame tube, out of harms way. The slides and needles are damaged very easily if this precaution is not observed.

2 Later models which have the push-pull throttle cable arrangement require disconnection of the throttle control cables only, from the link lever. In this instance, the carburettors are removed without need to disturb the throttle valve and needle assemblies at this stage.

3 Slacken off the clamp nearest to the carburettor on each flexible inlet stub and the clamp which retains each carburettor intake to the air cleaner box. Each series of clamps has a crosshead screw. It will be necessary to prise the intake stub from each carburettor because a lipped fitting is used. When the carburettors are free, they can be withdrawn as a complete in-line unit, still attached to their mounting plate.

4 To release each carburettor from the mounting plate, remove the two crosshead screws which pass through the mounting plate into the flange of the carburettor. In the case of carburettors fitted with the mechanical linkage, it will be necessary to lift off the rubber cap and slacken off the locknut at the top of each carburettor, so that the carburettor will be released from the forked end of the throttle linkage. DO NOT SLACKEN THE PINCH BOLT IN THE CLAMP AROUND THE OPERATING ROD, OR THE CARBURETTORS WILL NEED RESYNCHRONISING. Note that irrespective of the type of throttle control fitted, it will be necessary to detach the common choke control from each carburettor by withdrawing the split pin through the linkage connections. Mark each carburettor so that it is replaced in the original location.

6 Carburettors - dismantling, examination and reassembly

1 Invert each carburettor and remove the float chamber by prising the retaining clip to one side. The twin float assembly can be lifted away after the hinge pin has been displaced, giving access to the float needle and needle seating. Place the float needle in a safe place until reassembly commences; it is minute and very easily lost. The float needle seating will unscrew from the mixing chamber body.

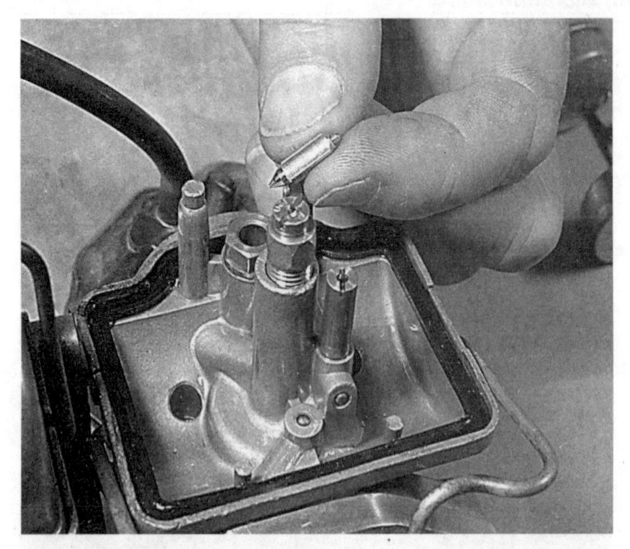

6.1a. Float needle will lift out of needle seating

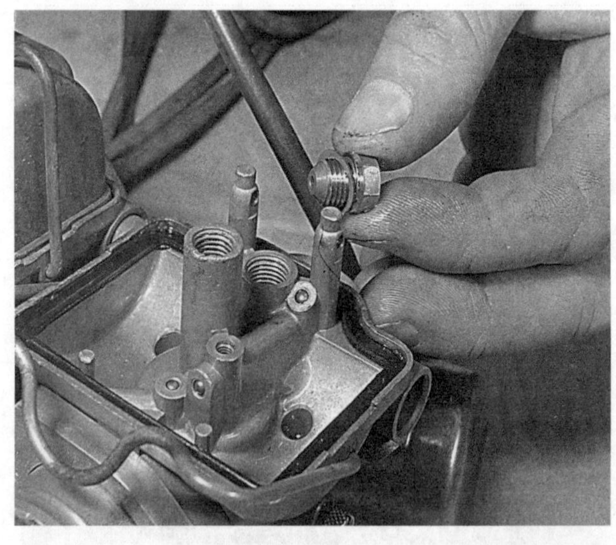

6.1b. Float needle seating will unscrew from carburettor body

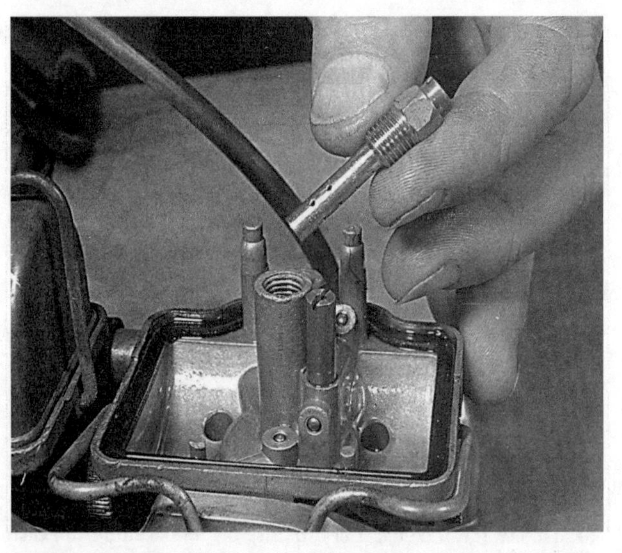

6.4. Main jet will probably unscrew together with needle jet

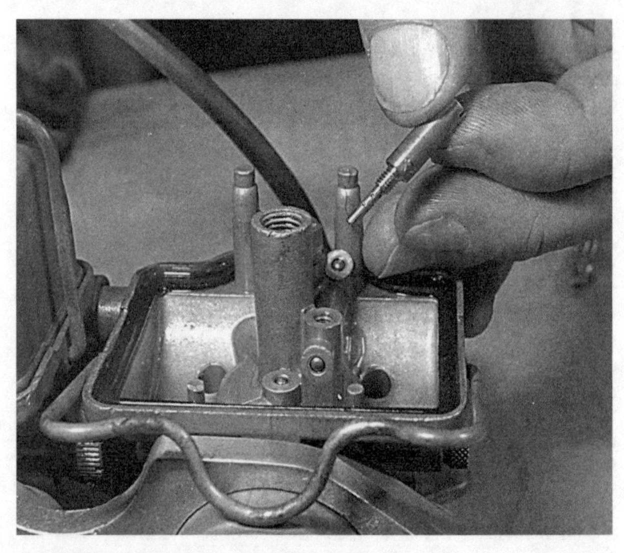

6.7. Pilot jet screws into carburettor body

6.8. Manually operated chokes extend into venturi when choke is applied

7.5. Throttle stop screw is fitted to individual carburettors on early models

8.2. This screw blanks off vacuum gauge attachment point

8.4. Pilot jet screw is found near carburettor intake

2 Note there is a gasket between the float chamber and the mixing chamber body which should not be disturbed unless it is damaged or has shown signs of leakage.

3 Check that the floats are in good condition and not punctured. Because they are made from brass, it is possible to effect a repair by soldering, but it is questionable whether such a repair can be justified other than in an emergency. The addition of solder will affect the weight of the float assembly and result in a different petrol level. Whenever possible, fit a new one.

4 The main jet is located in the centre of the oblong mixing chamber housing and will unscrew together with the needle. After a lengthy period of service, the needle and jet will wear and give rise to an increase in petrol consumption. It is advisable to renew the jet periodically, in conjunction with the needle itself.

5 The float needle seating will wear after lengthy service and should be closely examined with a magnifying glass. Wear usually takes the form of a ridge or groove, which will cause the float needle to seat imperfectly. Always renew the seating and float needle as a pair, especially since similar wear will almost certainly occur on the point of the needle.

6 The carburettor slides, or throttle valves, are attached to the cables which pass through the mixing chamber tops. The twist grip cable divides into four by means of a junction box normally hidden under the petrol tank, as a convenient means of actuating all four carburettors simultaneously. Later models, fitted with the push-pull cable linkage, use forked projections clamped around the linkage rod to actuate each carburettor via a threaded rod attached to each throttle valve. Remove the screwed top of the carburettor to gain access to the throttle valve assembly. Each slide has a return spring and the needle, suspended by means of a spring clip, from the centre. The needle is grooved, so that it can be either raised or lowered in order to vary the mixture.

7 The other jet threaded into the oblong mixing chamber is the pilot jet. This jet should be removed periodically and blown out with a jet of compressed air to ensure that the tiny air passages are not obstructed by any sediment from the petrol.

8 The manually-operated chokes are unlikely to require attention throughout the normal service life of the machine. When the operating plungers are depressed, flaps are lowered into the carburettor air intake which cut off the supply of air and therefore give a much richer mixture for cold starting. The machine should never be run for any distance with the chokes closed or the excessively rich mixture will foul the spark plugs and wash the oil from the cylinder walls, greatly accelerating the rate of engine wear.

9 Before the carburettors are reassembled, using the reversed dismantling procedure, each should be cleaned out thoroughly using compressed air. Avoid using a piece of rag since there is always risk of particles of lint obstructing the internal passageways or the jet orifices.

10 Never use a piece of wire or any pointed metal object to clear a blocked jet. It is only too easy to enlarge the jet under these circumstances and increase the rate of petrol consumption. If compressed air is not available, a blast of air from a tyre pump will usually suffice.

11 Do not use excessive force when reassembling a carburettor because it is easy to shear a jet or some of the smaller screws. Furthermore, the carburettors are cast in a zinc-based alloy which itself does not have a high tensile strength. Take particular care when replacing the mixing chamber tops to ensure the needles align with the jet seats and that the top is not engaged cross-threaded.

12 Avoid overtightening the screws which retain the carburettors to the mounting plate. Overtightening will cause the flanges to bow, giving rise to mysterious air leaks and a permanently weak mixture. If the flange is bowed, it can be rubbed down until it is flat once again using a rotary motion and a sheet of emery cloth wrapped around a sheet of glass. Make sure no particles of emery grit enter the carburettors and that the O ring in the centre of each flange is replaced when the grinding operation is complete.

13 Do NOT remove either the throttle stop screw or the pilot jet screw without first making note of their exact positions. Failure to observe this precaution will make it necessary to re-synchronise all four carburettors on reassembly. Note that the throttle stop screw has a large head clearly marked with graduation lines to aid setting adjustments. Later models have only a single throttle stop screw, on the linkage which actuates the carburettors.

7 Carburettors - adjustment

1 Before adjusting the carburettors, a check should be made to ensure that the following settings are correct: contact breaker gap, ignition timing, valve clearances, spark plug gaps, crankcase oil level. It is also important that the engine is at normal running temperature.

2 Remove the petrol tank and position it approximately 20 inches above the machine. Reconnect the tank to the carburettors by using longer fuel lines.

3 Remove the air cleaner upper and lower cases and check the operation of the choke flaps. They should open and close simultaneously. Check the clearance between the choke flap and the carburettor body. If greater than 0.020 inch (0.5 mm) the clearance should be adjusted to within this limit by lengthening the operating rod by means of the adjuster fitted to each carburettor linkage.

Fig. 2.2. Adjustment of carburettor operating rods

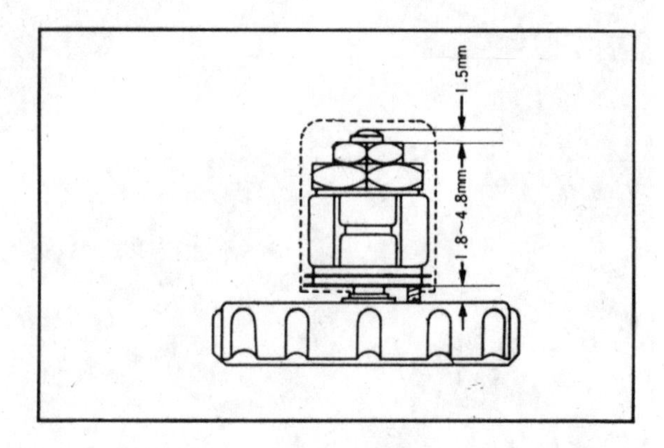

Fig. 2.3. Adjustment of carburettor operating rods

4 On models fitted with individual throttle cables, loosen the cable adjuster locknut on each carburettor top and adjust so that each throttle cable outer has free play within the range 0.040 - 0.080 inch (1 - 2 mm). Tighten all four locknuts and recheck the adjustment; all four cables should be adjusted uniformly. On later models, having the push-pull throttle cable linkage, there must be a clearance of 0.070 - 0.189 inch (1.8 - 4.8 mm) between the base of the adjuster screw and the top of each carburettor, with at least one thread showing above the uppermost locknut. If there is not, adjustment must be made, which will necessitate re-synchronising the carburettors. See the following Section.

5 Turn on the petrol and start the engine, allowing it to reach normal running temperature (oil temperature 140° - 157°F). Do NOT run the engine in a confined space or there is risk of carbon monoxide (CO) poisoning from the exhaust gases, which can have fatal results. If it is not possible to run the engine outside, good ventilation is essential. When the engine is at operating temperature, adjust the idling speed to 900 - 1000 rpm. This is accomplished by the throttle stop screw on models fitted with the push-pull throttle cable linkage. Turn the screw anticlockwise to increase the idling speed, clockwise to lower it. Models having the four separate throttle cables require adjustment of the throttle stop screw fitted to each carburettor, turning each only 1/8 of a turn at a time. It is difficult to arrive at a uniform setting for all carburettors and for this reason alone it is preferable to use a vacuum gauge so that each carburettor can be stabilised at a reading of 20 - 22 cm Hg.

6 Stop the engine and if there is no necessity to use a vacuum gauge or to synchronise the carburettors, replace the petrol tank and the original fuel lines, not forgetting the spring clips which retain them. Do not omit to replace the air cleaner cases.

8 Synchronising the carburettors

1 Power output will be unbalanced and slow running poor unless all four carburettors work in perfect harmony with each other. Many cases of poor overall performance and low power output can be attributed to carburettors which are out of phase with each other.

2 It is essential to use a vacuum gauge set comprising four separate dial gauges, one of each is connected to each carburettor by means of a special adaptor tube. The adaptor screws into the mounting flange on the right hand side of each carburettor, the orifice of which is normally blocked off by a cheese-head screw. Most owners are unlikely to possess the vacuum gauge set, which is somewhat expensive and is normally held by Honda agents who will carry out the entire synchronising operation for a nominal sum. It is questionable whether there is any advantage in purchasing a vacuum gauge set under these circumstances.

3 If the equipment is available and the carburettors are to be synchronised, it is important to check that the throttle operating rods on machines fitted with the push-pull cable linkage have at least one complete thread extending above the locknut. If there is not, remove the petrol tank and slacken off the throttle stop screw until there is slight clearance between the tip of the screw and the stop on which it abuts. Then slacken the locknut and adjust the adjuster screw until there is a clearance of 0.070 - 0.190 inch (1.8 - 4.8 mm) between the base of the adjuster screw and the top of the carburettor. Retighten the locknut and return the throttle stop screw to its original location. The carburettors must now be synchronised.

4 Raise the petrol tank to a level approximately 20 inches above the machine and use longer fuel lines to reconnect with the carburettors. Remove the cheese-head screws from each carburettor flange and screw in the adaptor so that a vacuum gauge is connected to each carburettor. Remove the upper and lower air cleaner cases. Start the engine and allow it to reach normal running temperature, observing the precautions about running the engine in a confined space, as detailed in the previous section. By either raising or lowering the adjuster screw in each carburettor top (push-pull cable linkage models only) arrange each vacuum gauge to read within the range 16 - 14 cm Hg, with no discrepancy greater than 3 cm Hg between ANY of the readings. Then adjust each individual pilot jet screw to ensure good engine rpm stability with maximum vacuum pressure. The pilot jet screw setting is usually from ¾ to 1¼ complete turns out from the fully closed position for optimum results. Finally, adjust the throttle stop screw until the engine idles smoothly in the 900 - 1000 rpm range and is not affected by snapping the throttle shut. Retighten all locknuts. Stop the engine, disconnect the vacuum gauges and adaptors, replace the cheese-head screws which seal off the adaptor connections, then replace the petrol tank and the original fuel lines, followed by the air cleaner cases.

5 A similar procedure is used for models having the four separate throttle cables. In this instance, the readings on the vacuum gauges are evened out by adjusting the individual throttle stop screws on each carburettor. The idling speed is set by adjusting the throttle stop screws an identical amount, varying only 1/8 turn at each adjusting stage whilst consulting the vacuum gauges. Turn clockwise to increase engine rpm, anticlockwise to lower it.

6 If needle swing on a vacuum gauge is too great, tighten the gauge restrictor valve so that swing is limited to within 2 cm Hg. If the indicated pressure is lower than 15 cm Hg, suspect an air leak or absence of slack in the throttle cable. Always pause after making an adjustment to the pilot jet screw to allow engine rpm to stabilise. Always refit the rubber caps after completing

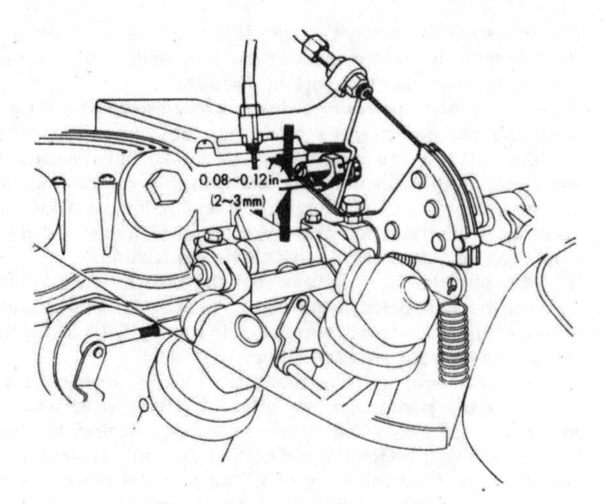

Fig. 2.4. Adjusting link pin clearance

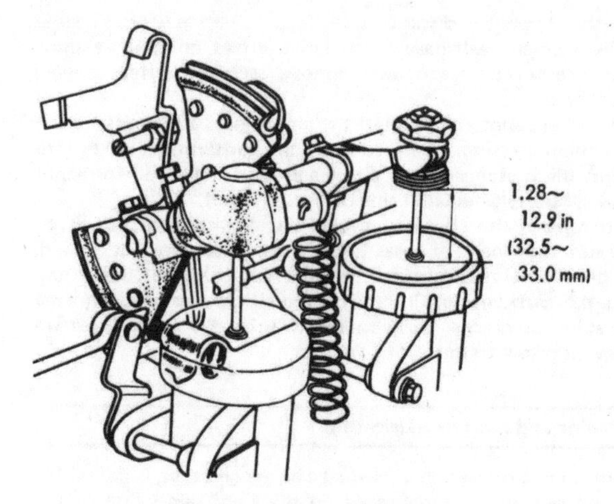

Fig. 2.5. Throttle stop adjustment

adjustments, making sure the rubber is not trapped by any part of the throttle movement.

9 Carburettor settings

1 Some of the carburettor settings, such as the sizes of the needle jets, main jets and needle positions, etc are pre-determined by the manufacturer. Under normal circumstances it is unlikely that these settings will require modification, even though there is provision made. If a change appears necessary, it can often be attributed to a developing engine fault.
2 As an approximate guide the pilot jet setting controls engine speed up to 1/8 throttle. The throttle slide cutaway controls engine speed from 1/8 to 1/4 throttle and the position of the needle in the slide from 1/4 to 3/4 throttle. The size of the main jet is responsible for engine speed at the final 3/4 to full throttle. It should be added however that these are only guide lines. There is no clearly defined demarkation line due to a certain amount of overlap which occurs between the carburettor components involved.
3 Always err slightly on the side of a rich mixture, since a weak mixture will cause the engine to overheat. Reference to Chapter 3 will show how the condition of the spark plugs can be interpreted with some experience as a reliable guide to carburettor mixture strength.

10 Carburettor cable adjustments

Push-pull cable linkage models only
1 Periodically check that there is clearance of from 0.080 - 0.12 inch (2 - 3 mm) between the throttle lever and the link pin (see accompanying diagram). The link pin is eccentrically mounted and it is necessary only to slacken the locknut and rotate the link pin in the appropriate direction to obtain the recommended clearance. Be sure to tighten the locknut after adjustment.
2 The clearance between the carburettor top and the base of the adjuster screw should be within the range 1.28 - 1.29 inch (32.5 - 33.0 mm) when the throttle is wide open. Adjustment of the setting is effected by the stop screw, as shown in the accompanying diagram.
3 There must be from 0.12 - 0.16 inch (3 - 4 mm) play in the handlebar twist grip before the carburettors commence to open. A cable adjuster is provided in the vicinity of the cable linkage to permit this adjustment to be made. Note that the throttle lever must hit the link pin when the twist grip is in the fully closed position. If it does not, the throttle cable must be renewed.

11 Exhaust system - cleaning

1 Unlike a two-stroke, the exhaust system does not require such frequent attention because the exhaust gases are usually of a less oily nature. However, instances may occur when it is necessary to withdraw the baffles from the silencers in order to clean them, especially if the engine is approaching the time for a rebore and is burning oil.
2 Each baffle tube assembly is retained by a single bolt, located in the side of the rear end of each silencer. When the screw is removed, the baffle tube can be withdrawn. There is no necessity to remove the exhaust system from the machine during this operation.
3 The baffles should be cleaned with a wire brush, or if the carbon is particularly oily, by washing with a petrol/paraffin mix. In an extreme case, a blow lamp can be used to burn off any heavy accumulation of carbon.
4 Do not run the machine with the exhaust baffles detached, or with a quite different type of silencer fitted. The standard production silencers have been designed to give the best possible performance, whilst subduing the exhaust note to an acceptable level. Although a modified exhaust system, or one without baffles, may give the illusion of greater speed as a result of the changed exhaust note, the chances are that performance will have suffered accordingly.
5 When replacing the baffles, make sure that the crosshead retaining screws are tightened fully. If they work loose, the baffle tubes may fall out whilst the machine is in motion.

12 Air cleaner - dismantling, servicing and reassembly

1 An air cleaner element, of corrugated paper is fitted across the frame immediately beneath the nose of the dualseat. Air enters through a cutaway in the rear of the box containing the element and through an angled flute in the front of the box. It passes through the element to each of the four carburettor intakes.
2 To gain access to the filter element, remove the lower case by unscrewing the two wing nuts located which pass upwards through the underside. The upper case is attached to the frame by a bolt which passes through a small lug on each end of the case and by the clamps which retain each carburettor intake hose. When the upper and lower cases are separated, the element can be lifted out.
3 To clean the element, tap it lightly to loosen the accumulation of dust and then use a soft brush to sweep the dust away. Alternatively, compressed air can be blown into the element from the inside. Remember the element is composed of corrugated paper and is easily damaged if handled roughly.

4 If the element is damp or oily, it must be renewed. A damp or oily element will have a restrictive effect on the breathing of the carburettor and will almost certainly affect engine performance.
5 On no account run without the air cleaners attached, or with the element missing. The jetting of the carburettors takes into account the presence of the air cleaner and engine performance will be seriously affected if this balance is upset.
6 To replace the element, reverse the dismantling procedure. Make sure that the inlet hoses to the carburettors are not kinked, split or otherwise damaged, otherwise air leaks will occur and affect the carburation. Check that the air cleaner cases are free from splits or cracks. Early models which have metal cases are somewhat prone to this.

13 Engine and gearbox lubrication

1 The engine oil, which is used also in the gearbox, is contained in a separate oil tank mounted on the right hand side of the machine, behind a detachable side cover. The tank has a capacity of 5.2 Imp pints (3.0 litres) and lubrication is effected on the dry sump principle.
2 Oil is fed by gravity to a trochoid mechanical pump, driven from kickstarter idler gear pinion. A spring-loaded valve shuts off the oil supply when the engine is stopped, to prevent oil draining into the crankcase via the pump. The delivery section of the pump feeds oil at a pre-set pressure via a pressure release valve, which by-passes the oil to the crankcase sump if the pressure exceeds the pre-set limit. As a result, it is possible to maintain constant pressure in the lubrication system. The standard setting is 56.9 psi at 4000 rpm engine speed, with an oil temperature of 80°C.
3 Since the oil flow will not, under normal circumstances, actuate the pressure release valve, it passes direct to the full flow oil filter which has a replaceable element, to filter out any impurities which may otherwise pass to the crankshaft assembly. The oil filter unit has its own by-pass valve to prevent the cut-off of the oil supply if the filter element becomes clogged.
4 Oil from the oil filter passes direct to the crankshaft and big end bearings, with a separate pressure feed to the overhead camshaft and rocker gear. A secondary flow passes to the gearbox via the gearbox main bearings. Surplus oil drains to the sump, where it is picked up by the scavenge section of the pump via a gauze strainer in the base of the pump casting. The scavenge pump returns the oil to the oil tank for re-circulation.
5 An oil pressure warning switch is included in the lubrication circuit to give visual warning by means of an indicator lamp if the pressure should fall to a low level.

14 Oil pump - dismantling, examination and reassembly

1 The oil pump can be removed and replaced with the engine in the frame. It is recommended, however, that the oil pump should not be removed and dismantled unless failure of the lubrication system has occurred or there is evidence of damage or a high rate of wear in the pump components. Unnecessary dismantling may give rise to additional problems, especially if the work is carried out under conditions which are anything but surgically clean.
2 Commence by removing the centre bolt from the oil filter housing at the front of the engine. Place a receptacle below the filter unit to catch the oil which will be released as the cover, complete with element, is removed. Remove the left hand side cover of the machine and drain the oil tank by unscrewing the drain plug. Approximately 5 pints of oil will be released.
3 Reposition the receptacle under the sump pan of the engine and remove the drain plug located at the end of the unfinned domed portion of the sump. Then remove the sump pan itself, which is retained by ten bolts. Access is now available to the oil pump, which is retained by three bolts. Unscrew these three bolts and withdraw the oil pump as a complete unit.
4 Transfer the pump to a clean working space before further dismantling takes place. The pump works with very small clearances and if particles of dirt find their way into the pump, its freedom of movement can be impaired, apart from the possibility of scoring the working surfaces.
5 Remove the three countersunk crosshead screws from the end cover of the pump and remove the rotor from the delivery section. Lift out the outer rotor, then take out the dowel pin which passes through the rotor shaft, so that the scavenge rotor assembly, complete with end cover and pinion, can be pulled away from the other end of the pump body. The inner and outer rotors can be pulled off the shaft in similar fashion.
6 Prise out the filter gauze from the base of the pump and unscrew the four bolts in the flanged base of the oil pump body to release the body itself from the main casting. Two of the bolts are located beneath the filter gauze.
7 The main casting contains the valve which shuts off the oil supply to the pump and the pressure release valve when the engine is stationary. The former is located behind the flange secured by two hexagon headed bolts and comprises a cylindrical plunger loaded by a light spring. The pressure release valve is located behind the hexagon headed end cap and is a somewhat similar plunger with a shouldered end, loaded by a much heavier gauge spring.
8 Wash all the pump components with petrol and allow them to dry before carrying out a full examination. Before part reassembling the pump for the various measurements to be made, check the castings for cracks or other damage, especially the pump end covers.
9 Reassemble the pump rotors and measure the clearance between the outer rotor and the pump body, using a feeler gauge. If the clearance exceeds 0.0138 inch (0.35 mm) the rotor or the body must be renewed, whichever is worn. Measure the clearance between the outer rotor and the inner rotor with a feeler gauge. If this clearance is greater than 0.0138 inch (0.35 mm) the rotors must be renewed as a set.
10 Examine the rotors and the pump body for signs of scoring, chipping or other surface damage which will occur if metallic particles find their way into the oil pump assembly. Renewal of the affected parts is the only remedy under these circumstances, bearing in mind that rotors must always be replaced as a matched set.
11 The clearance between the plunger of the oil flow stop valve and the body of the casting in which it slides, must not exceed 0.0067 inch (0.17 mm). The relief valve plunger should have a clearance which does not exceed 0.0039 inch (0.1 mm). Renewal of either the plunger or the main casting will be necessary if the wear limit is reached in either case.
12 Reassemble the pump and the pump casting by reversing the dismantling procedure. Make sure all parts of the pump are well lubricated before the end covers are replaced and that there is plenty of oil between the inner and outer rotors. Tighten the end covers down evenly and continually check that the drive pinion revolves freely up to the point where the crosshead screws have been tightened to their fullest extent. A stiff pump is usually due to dirt on the rotor faces. Before replacing the filter gauze in the base of the pump, ensure that it is clean.
13 Do not omit the O ring seals when the pump body is bolted to the main casting. Check that the cap of the oil pressure release valve is fully tightened. This also applies to the flange of the oil flow stop valve.
14 Replace the oil pump in the base of the crankcase, using new O rings at the oil passage joints. Fit a new gasket to the sump pan joint and refit the sump pan, using a light smearing of gasket cement to secure an oiltight joint. Do not omit to refit the drain plug in the sump pan, and in the oil tank, before refilling the latter.

15 Oil filter - renewing the element

1 As mentioned in the preceding Section, the oil filter element is contained within the cylindrical housing bolted to the front of the lower crankcase. To remove the housing, withdraw the

14.3. Remove sump pan for access to oil pump

14.3a. Pump is retained to crankcase by three bolts

14.5. Remove end cover for access to rotors

14.5a. Inner rotor will pull off drive shaft

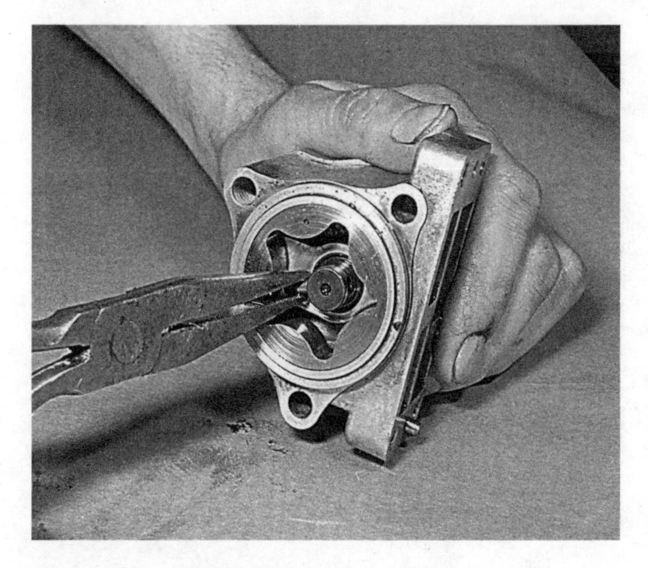

14.5b. Remove pin to free rotors on drive side

14.5c. Drive side inner rotor is retained by similar pin

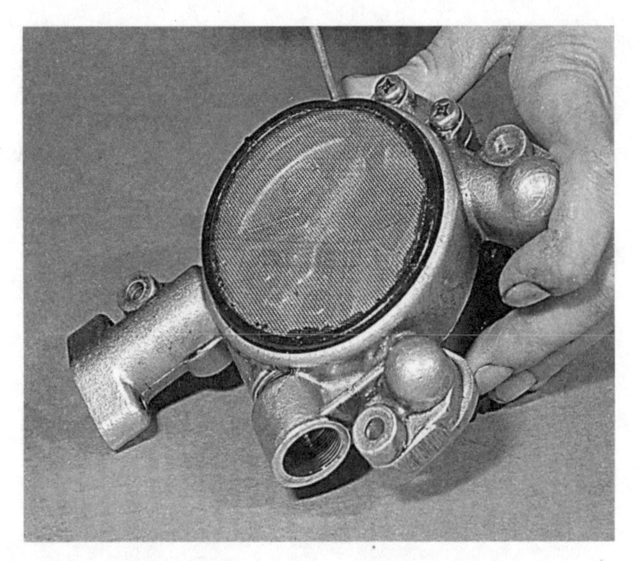

14.6. Filter gauze will prise from bottom of pump

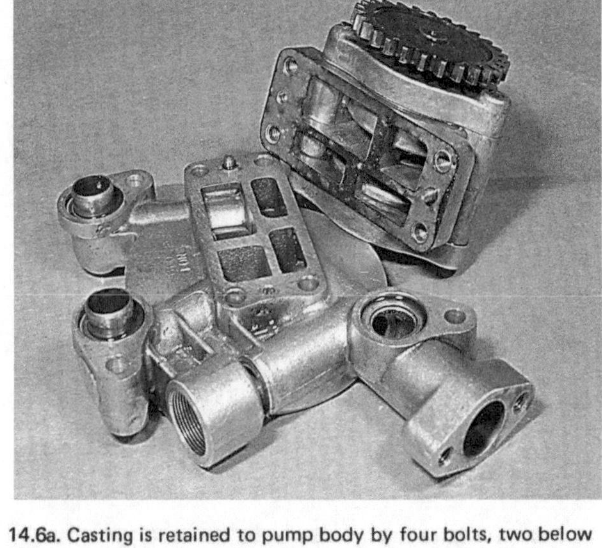

14.6a. Casting is retained to pump body by four bolts, two below filter gauze

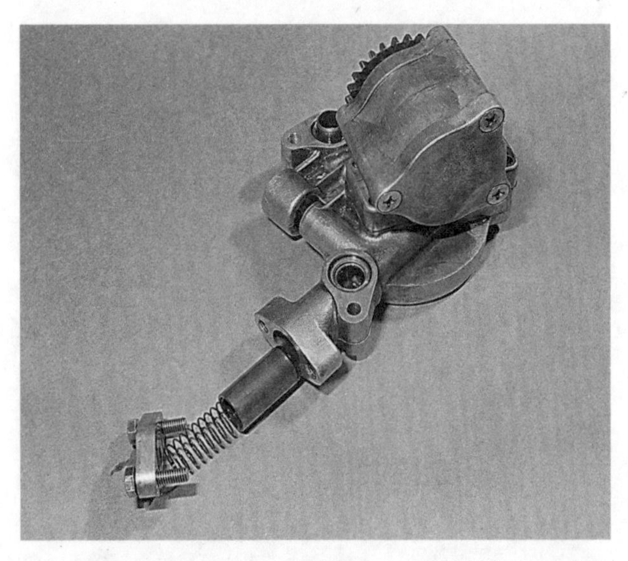

14.7. Oil shut off valve is located behind flange in pump casting

14.7a. Pressure release valve is behind hexagon-headed cap

14.9. Check clearance between outer rotor and pump body, also...

14.9a. ... clearance between inner and outer rotors

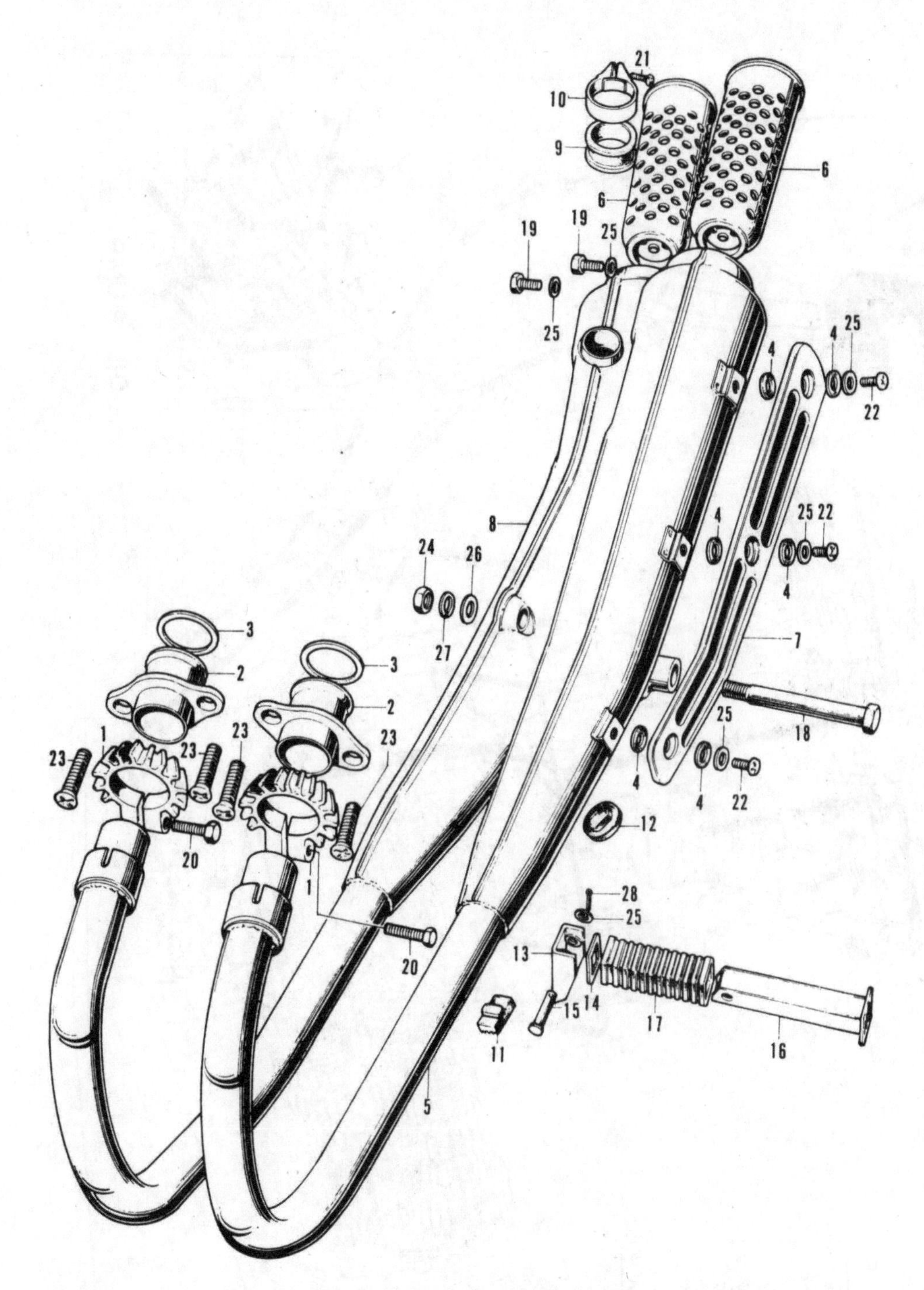

FIG. 2.6. EXHAUST PIPE SYSTEM

1	*Finned clamp - 4 off*
2	*Exhaust port stub - 4 off*
3	*Exhaust stub gasket - 4 off*
4	*Heat guard protector - 12 off*
5	*Exhaust pipe and silencer complete - 2 off **
6	*Exhaust diffuser - 4 off*
7	*Heat guard - 2 off*
8	*Exhaust pipe and silencer complete - 2 off **
9	*Silencer connecting tube - 2 off*
10	*Clamp - 2 off*
11	*Rear stand bump stop - 2 off*
12	*Prop stand stop*
13	*Pillion footrest bracket - 2 off*
14	*Pillion footrest washer - 2 off*
15	*Clevis pin for pillion footrest - 2 off*
16	*Pillion footrest bar - 2 off*
17	*Pillion footrest rubber - 2 off*
18	*Pillion footrest bolt - 2 off*
19	*Diffuser retaining bolt - 4 off*
20	*Finned clamp bolt - 4 off*
21	*Clamp screw - 2 off*
22	*Cross head screw - 6 off*
23	*Countersunk cross-head screws - 8 off*
24	*Pillion footrest retaining nut - 2 off*
25	*Plain washer - 12 off*
26	*Plain washer - 2 off*
27	*Spring washer - 2 off*
28	*Split pin - 2 off*
*	*Right-hand system is identical, but handed*

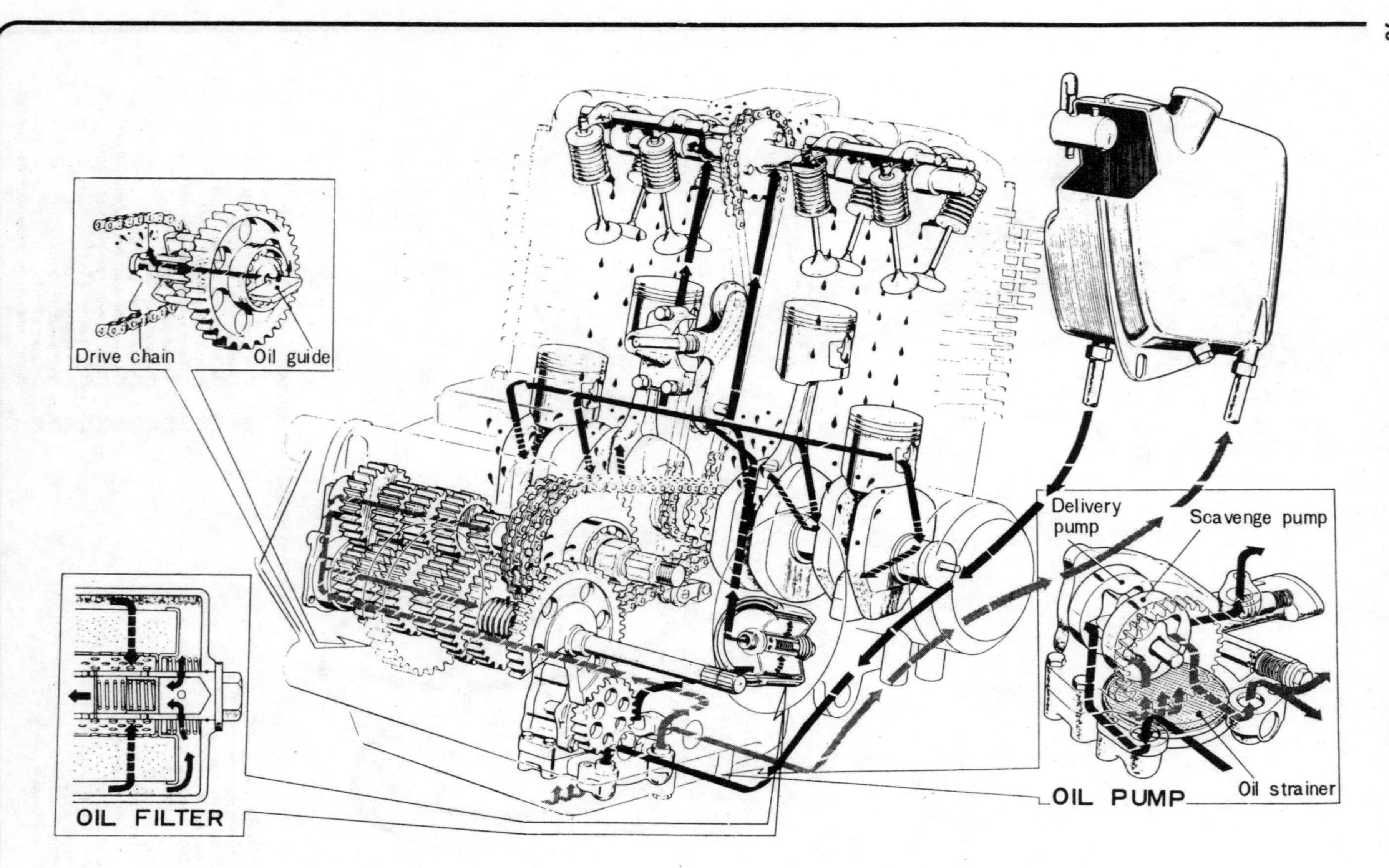

Fig. 2.7. Honda CB750 lubrication system

centre bolt, taking care to place a receptacle below the filter unit to catch the oil which will be released. The filter element can then be lifted off the housing and a new one fitted after cleaning out the old oil and sludge.

2 When renewing the filter element, it is advisable to renew the O ring seal around the main joint, at the same time. This will obviate the possibility of oil leakage. Do not overtighten the centre bolt when the filter housing is replaced.

3 Never run without the filter element or try to extend the period between oil changes or renewal of the element. Rapid wear of the engine components, especially the camshaft and rocker gear, will occur if this periodic routine maintenance task is neglected. Always check the filter unit for oil leaks after replacement, before setting out on a long run.

16 Oil pressure warning

1 An oil pressure switch is threaded into the top of the crankcase, to the rear of the cylinder block. It is connected to a red warning light in either the tachometer head or in an indicator lamp panel attached to the handlebars, to give visual warning if the oil pressure falls below a prescribed level. From a cold start, the lamp is normally extinguished when the oil pressure reaches 7.1 psi (0.5 kg cm).

2 If the oil warning lamp comes on whilst the machine is being ridden, the engine must be stopped immediately, otherwise there is risk of severe engine failure due to a breakdown of the lubrication system. The fault must be located and rectified before the engine is re-started and run even for a brief moment.

3 Cases of lubrication failure have occurred where particles of rubber have broken away from the inside of the flexible oil pipes, causing a blockage in the main feed. It is easy to check whether oil is being delivered to the oil pump by slackening the screws which retain the main feed pipe to the lower crankcase.

4 When the engine is operated at high temperatures, there may be a tendency for the oil warning lamp to come on occasionally, at idling speeds. This is quite in order if the light extinguishes immediately engine speed is increased.

5 Early models do not have armoured oil pipes and under certain circumstances a pipe will kink and cause low oil pressure. This is very difficult to detect because the pressure may not fall low enough to cause the warning lamp to light.

17 Primary drive lubrication

The primary drive chains do not require special attention with regard to lubrication because a chain oiler built into the drive shaft automatically fulfills this role. The twin chains are of the heavy duty type and because they run in ideal conditions, they should not require attention except after a very extensive mileage has been covered.

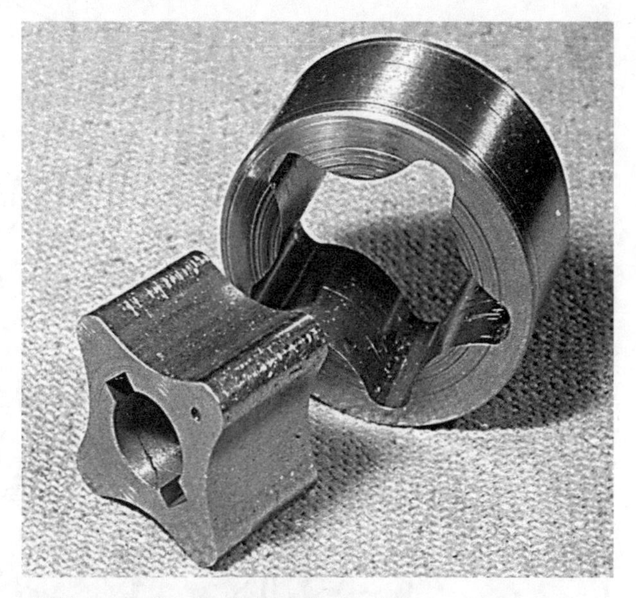

14.10. Examine rotor surfaces for blemishes

14.13. Do not omit 'O' rings when reassembling pump to crankcase

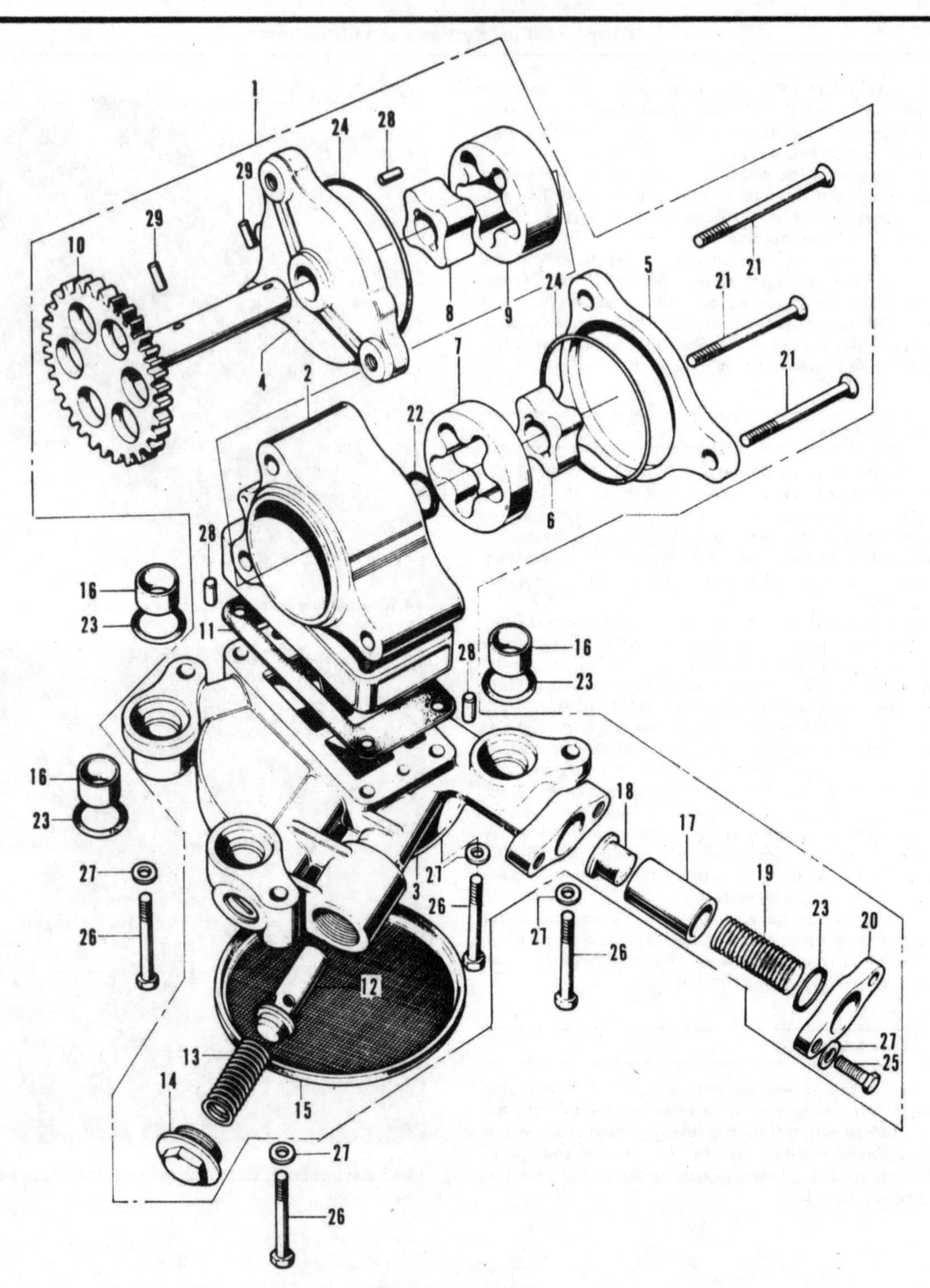

FIG. 2.8. OIL PUMP

1	*Oil pump assembly complete*	*16*	*'O' ring collar - 3 off*
2	*Oil pump body*	*17*	*Oil flow stop valve*
3	*Base casting*	*18*	*Oil flow stop seal*
4	*Right-hand cover*	*19*	*Oil flow stop spring*
5	*Left-hand cover*	*20*	*Oil flow stop spring cap*
6	*Inner rotor - delivery pump*	*21*	*Screw - 3 off*
7	*Outer rotor - delivery pump*	*22*	*Oil seal*
8	*Inner rotor - scavenge pump*	*23*	*'O' ring - 4 off*
9	*Outer rotor - scavenge pump*	*24*	*'O' ring - 2 off*
10	*Drive pinion*	*25*	*Bolt - 2 off*
11	*Oil pump gasket*	*26*	*Bolt - 7 off*
12	*Relief valve*	*27*	*Plain washer - 7 off*
13	*Relief valve spring*	*28*	*Dowel pin - 3 off*
14	*Relief valve spring cap*	*29*	*Rotor retaining pin - 2 off*
15	*Oil strainer*		

18 Fault diagnosis

Symptom	Reason/s	Remedy
Engine gradually fades and stops	Fuel starvation	Check vent hole in filler cap. Sediment in filter bowl or float chamber. Dismantle and clean.
Engine runs badly. Black smoke from exhausts	Carburettor flooding	Dismantle and clean carburettor. Check for punctured float or sticking float needle.
Engine lacks response and overheats	Weak mixture Air cleaner disconnected or hose split Modified silencer has upset carburation	Check for partial block in carburettors. Reconnect or renew hose. Replace with original design.
Oil pressure warning light comes on	Lubrication system failure	Stop engine immediately. Trace and rectify fault before re-starting.
Engine gets noisy	Failure to change engine oil when recommended	Drain off old oil and refill with new oil of correct grade. Renew oil filter element.

Chapter 3 Ignition system

Contents

Specifications

Alternator

Type	Excited field, 3 phase
Make	Hitachi
Output	12 volts, 13 amps, AC
Polarity	Negative earth
Stator coil resistance	0.2 ohms
Field coil resistance	7.2 ohms

Contact breaker gaps	0.012 - 0.016 inch (0.3 - 0.4 mm)
Ignition timing	6° BTDC @ 1000 rpm, 40° BTDC @ 2500 rpm Verified by timing marks

Ignition coils

Make	Toyo Denso

Condensers

Capacity	0.24 mf ± 10%
Insulation resistance	Over 10M ohms

Spark plugs

Make	NGK	Nippon Denso
Type	D8ES	X27ES-U
Size	12 mm	
Reach	19 mm (¾ in)	
Gap	0.024 - 0.028 in (0.6 - 0.7 mm)	

1 General description

The spark necessary to ignite the petrol/air mixture in the combustion chambers is derived from a battery and twin ignition coils, one for each pair of cylinders. A double contact breaker operates in conjunction with a twin lobe cam to determine the exact moment at which the spark will occur in each cylinder. When the contact breaker points separate the low tension circuit is interrupted and a high tension voltage is developed by the coil. This jumps the air gap across the points of the spark plug due to fire and ignites the mixture.

The AC generator (alternator) attached to the extreme left hand end of the crankshaft generates an alternating current which is rectified and used to charge the battery. Output is controlled by a dual contact electro-mechanical regulator which keeps the charge rate in accordance with the rate of electrical demand on the battery. A rectifier and regulator unit are housed in the electrical compartment under the dualseat in close proximity to the battery, and the twin ignition coils are mounted immediately behind the steering head, covered by the petrol tank. A contact breaker assembly is attached to the extreme right hand end of the crankshaft, within a separate compartment sealed with a chromium plated inspection cover. The auto-advance mechanism is housed behind the contact breaker baseplate and the circuit is actuated by a switch immediately below the nose of the petrol tank, on the left hand side of the machine. It operates the ignition circuit and the lights when running or parking.

2 AC generator - checking the output

1 Before checking the output from the generator, first ensure that the battery is fully charged. The voltmeter reading should be not less than 12 volts; if the check is carried out immediately after charging, the voltage may be as high as 15 - 16 volts on open circuit.

2 Disconnect the red and red/white leads from the battery positive (+) terminal and connect both to the positive (+) terminal of a 0 - 10 amp ammeter. Connect a short length of insulated wire between the negative (—) terminal of the ammeter and the positive (+) terminal of the battery. So that the battery voltage can be monitored throughout the check, connect a dc voltmeter across the battery terminals; positive lead to battery positive (+) terminal and negative lead to battery negative (—) terminal.

3 Start the engine and observe the ammeter readings with the main switch in the daytime riding position and then when the headlamp is on main beam. Do not operate the turn signal or the stop lamp during these tests. Operate the engine at different speeds and relate the ammeter reading to engine rpm. The following table shows the readings which should be obtainable; if they are lower the regulator unit may require adjustment, as described in Section 6 of Chapter 6. If there is no charge or the readings are unduly low, seek the aid of either a Honda agent or an auto-electrical engineer.

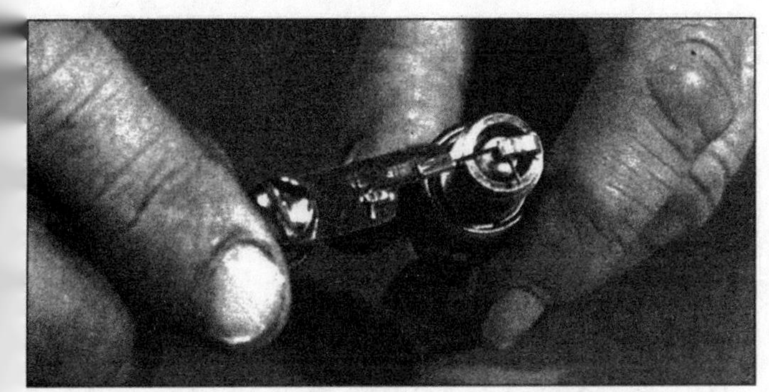

Electrode gap check - use a wire type gauge for best results

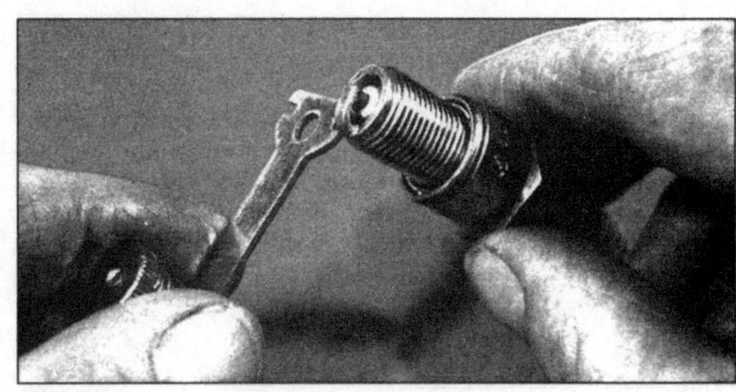

Electrode gap adjustment - bend the side electrode using the correct tool

Normal condition - A brown, tan or grey firing end indicates that the engine is in good condition and that the plug type is correct

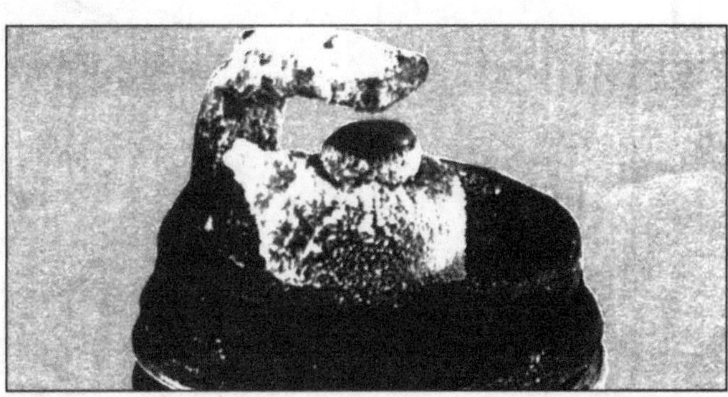

Ash deposits - Light brown deposits encrusted on the electrodes and insulator, leading to misfire and hesitation. Caused by excessive amounts of oil in the combustion chamber or poor quality fuel/oil

Carbon fouling - Dry, black sooty deposits leading to misfire and weak spark. Caused by an over-rich fuel/air mixture, faulty choke operation or blocked air filter

Oil fouling - Wet oily deposits leading to misfire and weak spark. Caused by oil leakage past piston rings or valve guides (4-stroke engine), or excess lubricant (2-stroke engine)

Overheating - A blistered white insulator and glazed electrodes. Caused by ignition system fault, incorrect fuel, or cooling system fault

Worn plug - Worn electrodes will cause poor starting in damp or cold weather and will also waste fuel

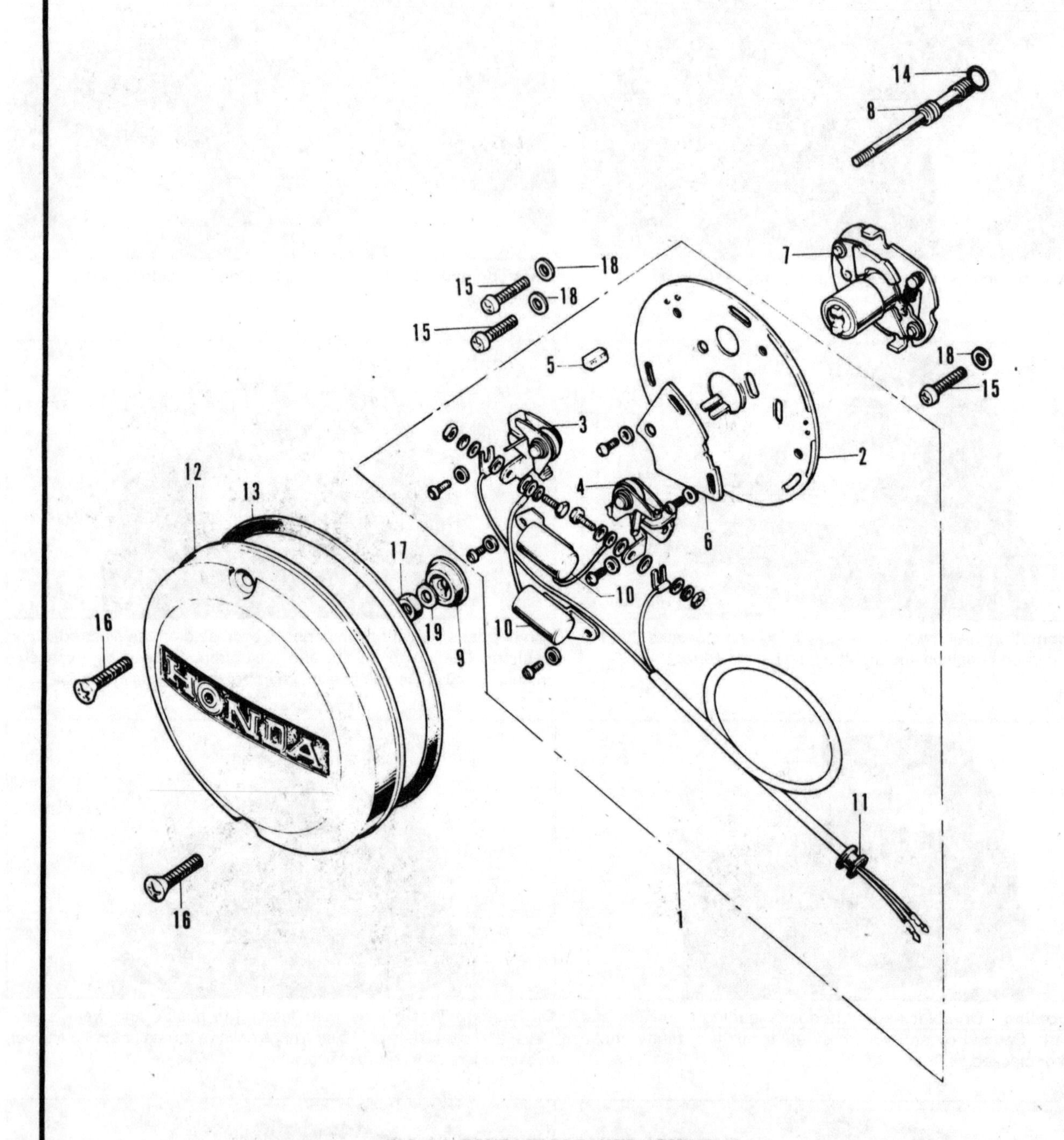

FIG. 3.1. CONTACT BREAKER ASSEMBLY

1	Contact breaker assembly (Hitachi)	11	Grommet
2	Contact breaker plate	12	Contact breaker cover
3	Right-hand contact breaker	13	Gasket for contact breaker cover
4	Left-hand contact breaker	14	'O' ring
5	Oil felt	15	Cross head screws - 3 off
6	Contact breaker moving plate	16	Contact breaker cover screw - 2 off
7	Auto-advance assembly and contact breaker cam	17	Nut
8	Auto-advance shaft	18	Plain washer - 3 off
9	Special washer for shaft	19	Plain washer
10	Condenser		

Engine rpm	1000	2000	3000	4000	5000	6000	7000	8000
				Charging current (amps)				
Day riding	2–3	1	1	1	1	1	1	1
Night riding	6.5	0	2.4	1.3	1.0	1.0	0.8	0.6
Battery terminal voltage	12	12.4	13.2	14.5	14.5	14.5	14.5	14.5

3 Ignition coils - checking

1 Each ignition coil is a sealed unit, designed to give long service without need for attention. It is located within the top frame tubes, immediately to the rear of the steering head assembly. If a weak spark and difficult starting causes the performance of a coil to be suspect, it should be tested by a Honda agent or an auto-electrical engineer who will have the appropriate test equipment. A faulty coil must be renewed;it is not possible to effect a satisfactory repair.
2 A defective condenser in the contact breaker circuit can give the illusion of a defective coil and for this reason it is advisable to investigate the condition of the condenser before condemning the ignition coil. Refer to Section 6 of this Chapter for the appropriate details.
3 Note that it is extremely unlikely that both ignition coils will prove faulty at the same time, unless the common electrical feed is in some way derranged. This can be checked by measuring the low tension voltage supplied to the coils, using a voltmeter.

4 Contact breaker - adjustments

1 To gain access to the contact breaker assembly, it is necessary to detach the chromium plated cover retained by two crosshead screws at the right hand end of the crankshaft. Note that the cover has a sealing gasket, to prevent the ingress of water.
2 Rotate the engine slowly by the kickstarter until one set of points is in the fully open position. Examine the faces of the contacts. If they are blackened and burnt, or badly pitted, it will be necessary to remove them for further attention. See Section 5 of this Chapter. Repeat for the second set of contact points.
3 Adjustment is effected by slackening the screw through the plate of the fixed contact breaker point and moving the point either closer to or further from the moving contact until the gap is correct as measured by a feeler gauge. The correct gap with the points FULLY OPEN is 0.012 - 0.016 inch (0.3 - 0.4 mm). Small projections on the contact breaker baseplate permit the insertion of a screwdriver to lever the adjustable point into its correct location. Repeat this operation for the second set of points, which must also be fully open.
4 Do NOT slacken the two screws through the extremities of the larger baseplate fitted to the right hand set of contact breaker points. They are used for adjusting the setting of the ignition timing and it will be necessary to re-time the engine if the baseplate is permitted to move. Only the centre screw should be slackened, to adjust the fixed contact breaker point.
5 Before replacing the cover and gasket, space a light smear of grease on the contact breaker cam and one or two drops of thin oil on the left which lubricates the surface of the cam. It is better to under-lubricate rather than add excess because there is always chance of excess oil reaching the contact breaker points and causing the ignition circuit to malfunction.

5 Contact breaker points - removal, renovation and replacement

1 If the contact breaker points are burned, pitted or badly worn, they should be removed for dressing. If it is necessary to remove a substantial amount of material before the faces can be restored, the points should be renewed.
2 To remove the contact breaker points, detach the circlip which secures the moving contact to the pin on which it pivots. Remove the nut and bolt which secures the flexible lead wire to the end of contact return spring, noting the arrangement of the insulating washers so that they are replaced in their correct order during reassembly. Lift the moving contact off the pivot, away from the assembly.
3 The fixed contact is removed by unscrewing the two screws which retain the contact to the contact breaker baseplate.
4 The points should be dressed with an oilstone or fine emery cloth. Keep them absolutely square throughout the dressing operation, otherwise they will make angular contact on re-assembly, and rapidly burn away.
5 Replace the contacts by reversing the dismantling procedure, making sure that the insulating washers are fitted in the correct order. It is advantageous to apply a thin smear of grease to the pivot pin, prior to replacement of the moving contact arm.
6 Check, and if necessary, re-adjust the contact breaker gap when the points are fully open. Repeat the whole operation for the second set of points.

6 Condensers - removal and replacement

1 A condenser is included in each contact breaker circuit to prevent arcing across the contact breaker points as they separate. It is connected in parallel with each set of points and if a fault develops, ignition failure is liable to occur.
2 If the engine proves difficult to start, or misfiring occurs, it is possible that the condenser is at fault. To check, separate the contact breaker points by hand when the ignition is switched on. If a spark occurs across the points and they have a blackened and burnt appearance, the condenser can be regarded as unserviceable.
3 It is not possible to check a condenser without the appropriate test equipment. In view of the low cost involved, it is preferable to fit a new one and observe the effect on engine performance.
4 Because each condenser and its associated set of contact breaker points is common to a pair of cylinders, a faulty condenser will not cause a misfire on one cylinder only. In such a case it is necessary to seek the cause of the trouble elsewhere, possibly in some other part of the ignition circuit or the carburettor. It also follows that both condensers are unlikely to fail at the same time unless damaged in an accident. If the cases are crushed or dented, electrical breakdown will occur.
5 The condensers are located at the base of the contact breaker assembly, parallel to each other. Each has an integral bracket and is attached to the contact breaker baseplate by a single crosshead screw, making renewal easy.

7 Ignition timing - checking and resetting

1 In order to check the accuracy of the ignition timing, it is

necessary to remove the contact breaker cover at the extreme right hand end of the crankshaft, by detaching the two crosshead screws. The cover will lift away, complete with the sealing gasket. There is a circular aperture in the contact breaker baseplate and the base of the auto-advance unit which lies behind the contact breaker assembly is inscribed with two sets of lines at 180° to each other, marked 'T' and 'F'. There is also a fixed reference point in the form of an aluminium alloy projection, with a timing mark formed in the centre. The auto-advance unit has a dowelled fitting in the end of the crankshaft and can be replaced in one fixed position only. In consequence, the timing marks cannot be varied in relationship to one another.

2 If the ignition timing is correct, the 'F' line will coincide exactly with the static timing mark when the contact breaker points relating to the cylinder at firing point are about to separate. The figures marked on the base of the rotating auto-advance unit will identify the cylinders concerned, either 1 and 4 or 2 and 3. Both sets of points must be checked in this manner.

3 To adjust the position of the contact breaker points in relation to the cam, slacken the three crosshead screws around the periphery of the contact breaker baseplate. The plate is slotted to permit a limited amount of movement; move the plate anticlockwise to advance the timing or clockwise if it has to be retarded. Note that this adjustment will affect BOTH pairs of cylinders.

4 If the timing is accurate on one pair of cylinders and not the other, arrange it so that the timing is correct when the left hand set of points are in operation. The right hand set of points have an extra wide mounting plate which permits a limited amount of adjustment on this set of points alone. Slacken the two screws at each end of the base mounting, obtain the correct setting of the points in relation to the cam and then retighten the screws. The points must be about to separate when the timing marks exactly coincide.

5 Greater accuracy of the setting is achieved if a bulb is used to indicate when separation of the points occurs. Disconnect the electrical leads from the set of points being checked and attach a lead from one terminal of the bulb in its place. Connect the other terminal of a bulb to the positive terminal of a battery and the negative lead of the battery to the crankcase. When the points separate, the electrical circuit will be broken and the lamp will extinguish. A cycle lamp battery and bulb are quite suitable for this operation.

6 It cannot be overstressed that optimum performance depends on the accuracy with which the ignition timing is set. Even a small error can cause a marked reduction in performance and in an extreme case, engine damage as the result of overheating. The contact breaker gaps must be checked and if necessary reset to the recommended clearance BEFORE the accuracy of the ignition timing is verified. Adjustments to the gaps made afterwards will affect the accuracy of the setting. Do NOT confuse the 'F' mark with the adjacent 'T' mark on the rotating base of the auto-advance unit. The 'T' mark indicates when the pistons in each pair of cylinders are at top dead centre (TDC).

7 Although the base of the auto-advance unit carries the timing markings, the ignition is timed when the unit is in the static position (balance weights unextended).

8 Automatic ignition advance unit - examination

1 The automatic ignition advance mechanism rarely requires attention, although it is advisable to examine it periodically, when the contact breaker is receiving attention. It is retained by a small bolt and washer through the centre of the integral contact breaker cam and can be pulled off the end of the camshaft when the contact breaker plate is removed.

2 The unit comprises spring loaded balance weights, which move outward against the spring tension as centrifugal force increases. The balance weights must move freely on their pivots and be rust-free. The tension springs must also be in good condition. Keep the pivots lubricated and make sure the balance weights move easily, without binding. Most problems arise as a

3.1. Ignition coils are located to rear of steering head

4.3. Small projections aid screwdriver adjustment of fixed point

4.4. Screws in extremities of base plate permit movement of right-hand points, independent of whole contact breaker unit

result of condensation within the engine, which causes the unit to rust and balance weight movement to be restricted.
3 The auto-advance mechanism is fixed in relation to the crankshaft by means of a dowel. In consequence the mechanism cannot be replaced in anything other than the correct position. This ensures accuracy of ignition timing to within close limits, although a check should always be made when reassembly of the contact breakers is complete.

9 Spark plugs - checking and resetting the gaps

1 A matched set of 12 mm medium reach spark plugs are fitted to the Honda CB 750 4's. The choice of plug grade depends on the operating conditions. For normal use, NGK plugs of the D-8ES type are recommended or the equivalent Denso X-24ES. For high speed work a colder running plug of the NGK D-10E is recommended, or if traffic work is involved in the lower speed range, a hotter running plug of the NGK type D-7ES. Do not deviate from any of these recommended grades; they will give the best all round results.
2 All spark plugs fitted to the Honda CB 750, irrespective of grade, must be gapped within the range 0.024 - 0.028 inch (0.6 - 0.7 mm). This applies also to the British-made spark plugs listed as alternatives in the Specifications Section of this Chapter.
3 Check the gap of the plug points during every 3 monthly or 2000 mile service. To reset the gap, bend the outer electrode to bring it closer to the centre electrode and check that a 0.6 mm (0.024 inch) feeler gauge can be inserted. Never bend the central electrode or the insulator will crack, causing engine damage if particles fall in whilst the engine is running.
4 With some experience, the condition of the spark plug electrodes and insulator can be used as a reliable guide to engine operating conditions. See the accompanying diagram.
5 Always carry a spare pair of spark plugs of the recommended grade. In the rare event of plug failure, they will enable the engine to be restarted.
6 Beware of over-tightening the spark plugs, otherwise there is risk of stripping the threads from the aluminium alloy cylinder heads. The plugs should be sufficiently tight to seat firmly on their copper sealing washers, and no more. Use a spanner which is a good fit to prevent the spanner from slipping and breaking the insulator.
7 If the threads in the cylinder head strip as a result of over-tightening the spark plugs, it is possible to reclaim the head by the use of a Helicoil thread insert. This is a cheap and convenient method of replacing the threads; most motor cycle dealers operate a service of this nature at an economic price.
9 Make sure the plug insulating caps are a good fit and have their rubber seals. They should also be kept clean to prevent tracking. These caps contain the suppressors that eliminate both radio and TV interference.

10 Fault diagnosis

Symptom	Reason/s	Remedy
Engine will not start	Faulty ignition switch	Operate switch several times in case contacts are dirty. If lights and other electric function, switch may need renewal.
	Starter motor not working	Discharged battery. Use kickstarter until battery is recharged.
	Short circuit in wiring	Check whether fuse is intact. Eliminate fault before switching on again.
	Completely discharged battery	If lights do not work, remove battery and recharge.
Engine misfires	Faulty condenser in ignition circuit	Replace condenser and re-test.
	Fouled spark plug	Replace plug and have original cleaned.
	Poor spark due to generator failure and discharged battery	Check output from generator. Remove and recharge battery.
Engine lacks power and overheats	Retarded ignition timing	Check timing and also contact breaker gap. Check whether auto-advance mechanism has jammed.
Engine 'fades' when under load	Pre-ignition	Check grade of plugs fitted; use recommended grades only.

Chapter 4 Frame and Forks

Contents

Specifications

Front forks

Fork springs - free length	19.075 in (484.5 mm)	
- serviceable limit	18.11 in (460 mm)	
Fork piston - outer diameter	1.552 - 1.553 in (39.425 - 39.45 mm)	
- serviceable limit	1.5512 in (39.4 mm)	
Lower fork leg - internal diameter	1.555 - 1.556 in (39.50 - 39.534 mm)	
- serviceable limit	1.5591 in (39.68 mm)	
Damping fluid oil type	SAE 10W/30	
Damping fluid capacity	**After draining**	**Dry**
CB750, K1 and K2	200 - 210 cc	220 - 230 cc
CB750 K3 to K5	130 - 135 cc	150 - 155 cc
CB750 K6	135 - 145 cc	155 - 160 cc
CB750 F	125 - 135 cc	145 - 155 cc

Rear suspension

Pivot bush - inner diameter	0.8444 - 0.8456 in (21.448 - 21.5 mm)
- serviceable limit	0.8543 in (21.7 mm)
Suspension units (hydraulic type)	
free length of spring	8.74 in (222 mm)
serviceable limit	8.504 in (216 mm)
Suspension units (gas-filled type)	
free length of spring	8.58 in (218 mm)
serviceable limit	8.346 in (212 mm)
Overall length of suspension units (all types)	13.07 in (332 mm)
Stroke	3.4 in (87 mm)

1 General description

All Honda CB 750's have a duplex tube frame of the full cradle type, fitted with an exceptionally strong steering head designed to provide the degree of rigidity essential for high speed riding. Rear suspension is of the swinging arm type, using units either gas or oil-filled, to provide the necessary damping action. The units are adjustable so that the spring ratings can be varied within certain limits to match the load carried.

The front forks are of the conventional telescopic type, having internal oil-filled dampers. The fork springs are contained within the fork stanchions and each fork leg can be detached from the machine as a complete unit, without dismantling the steering head bearings.

2 Front forks - removal from the frame

1 It is unlikely that the front forks will have to be removed from the frame as a complete unit, unless the steering head assembly requires attention, or if the machine suffers frontal damage.

2 To detach the complete fork assembly, commence by removing the master cylinder body of the hydraulic front brake which is secured to the handlebars by two bolts. Then disconnect the clutch cable from the clutch lever.

3 Remove the combined starter, lighting and ignition switch assembly on the right hand end of the handlebars AFTER first disconnecting the positive lead from the battery. It is retained by two screws on the underside. Detach the twist grip cable.

4 Detach the front of the headlamp and disconnect the electrical wiring at the snap connectors. The wires leading to the handlebars can be pulled through the orifice in the rear of the headlamp, complete with their connectors.

5 Remove the four bolts which retain the two handlebar split clamps in position, and lift the handlebars away. Slacken the clamp which retains the tachometer and speedometer heads and remove these instruments from the fork top yoke.

6 Slacken the pinch bolts at the top of the upper yoke, to release the grip on the individual fork legs. Remove the domed nut in the centre of the fork stem and the pinch bolt at the rear of the upper yoke. The yoke can then be lifted off the top of the fork assembly.

7 Place a sturdy support under the engine unit so that the front wheel is raised clear of the ground. Detach the speedometer cable from the hub of the front wheel; it is retained by single

2.7. Speedometer drive cable is retained by a single screw

2.7a. Front wheel spindle is freed by releasing split clamps

2;7b. Wheel will lift away complete with brake disc

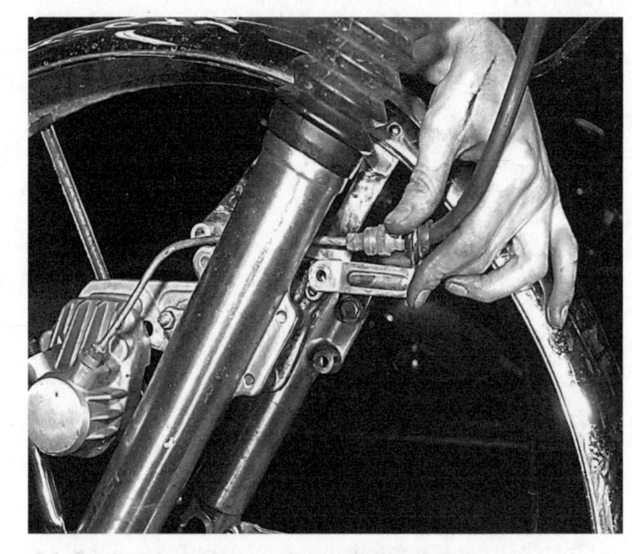

2.9. Single bolt retains support for hydraulic hose

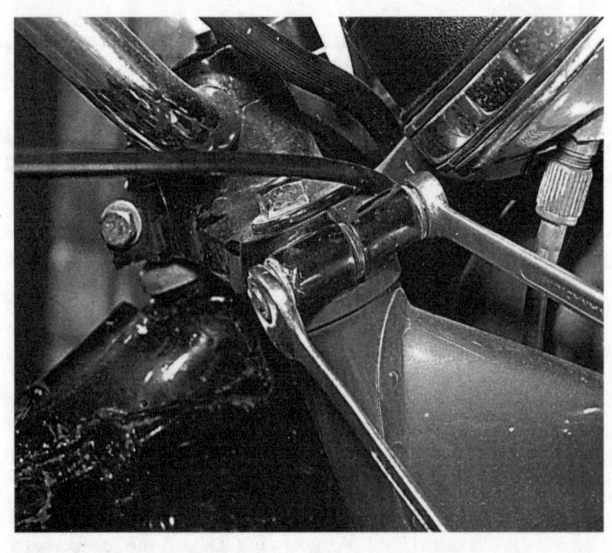

2.10. Slacken the pinch bolts through the upper fork yoke

2.11. Pinch bolts through the lower yoke must also be slackened

crosshead screw, which must be removed. Remove the nuts and washers from the lower ends of each fork leg, so that the split clamps which retain the wheel spindle can be pulled free. The wheel is now released from the fork ends, complete with spindle.
8 Remove the three bolts which retain the hydraulic front brake caliper and also the adjuster nut, making a note of the setting of the latter. The three bolts are located on the inside of the lower left hand fork leg and the adjuster nut on the outside, the latter controlling the amount of swing of the caliper. The complete caliper unit is then freed from the lower left hand fork leg without need to disturb the hydraulic system.
9 Remove the small bolt which retains the support clip for the hydraulic hose and pipe assembly. It is found at the rear of the lower left hand fork leg, close to the mudguard stays. Make arrangements to support the caliper and hose assembly whilst it is detached from the fork leg, so that the strain is not taken by hose or pipe. Beware of kinking either the hose or the pipe and under no circumstances operate the front brake or there is risk of forcing the caliper piston out of the cylinder, with the resulting loss of hydraulic fluid. Should this happen, the brake assembly must be bled after it has been replaced and the front wheel is in position.
10 Remove the front mudguard, which is retained by two bolts at the lower end of each fork leg, and by two additional bolts on the inside of each fork leg. Remove the mudguard complete with the centre bridge and stays. Unscrew the pinch bolts through each side of the upper fork yoke.
11 Unscrew the pinch bolts through the lower fork yoke, detach the rubber gaiters from the upper fork shrouds, and pull the individual fork legs downward so that each is released from the yoke as a complete unit. It may be necessary to spring the clamp apart with a screwdriver blade to release the grip on the fork stanchions. Note that the headlamp shell and upper fork shrouds will remain in position and can be lifted off when the fork legs are withdrawn.
12 To release the lower yoke and steering head stem, unscrew the slotted sleeve nut at the top of the steering head column, using either Honda service tool 07072-20002, a 'C' spanner or a punch. As the steering head column is released, the uncaged ball bearings from the lower race will be released and care should be taken to catch them by wrapping a rag around the bearing area. The bearings in the upper race will almost certainly remain in position.
13 It follows that much of this procedure can be avoided if it is necessary to remove the individual fork legs without disturbing the steering head assembly or bearings. Under the circumstances, commence dismantling as described in paragraphs 7 to 10, then slacken the pinch bolts in the upper and lower fork yokes, so that the fork legs can be withdrawn as complete units.

2.11a. If necessary, spring clamp apart whilst pulling fork leg downwards

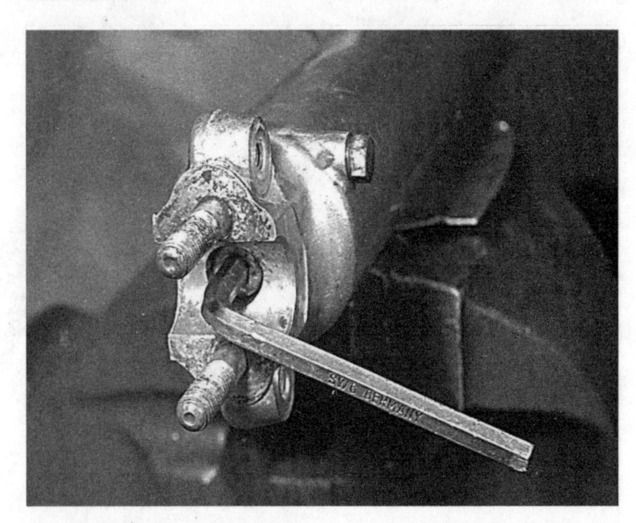

3.2. Socket screw is recessed into bottom of lower fork leg

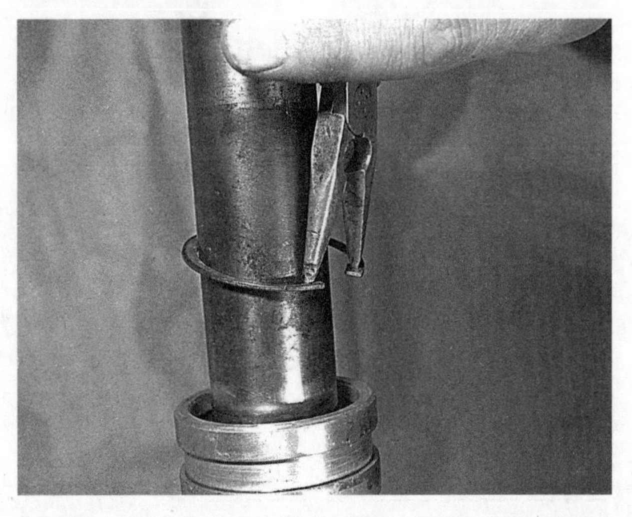

3.2a. Remove circlip to detach stanchion tube from lower fork leg

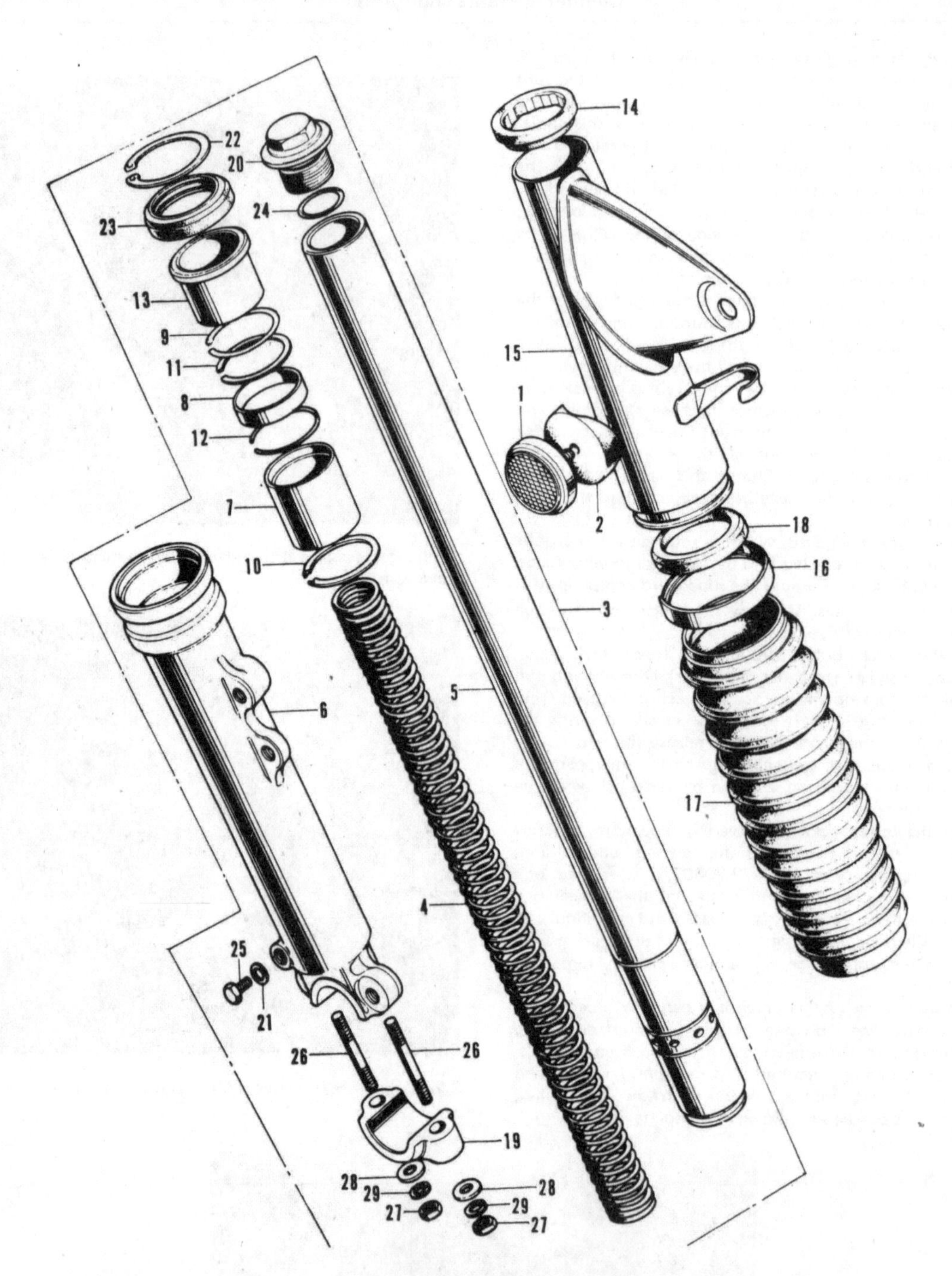

FIG. 4.1. TELESCOPIC FORKS

1	*Reflector unit*	*16*	*Cap for shroud - 2 off*
2	*Reflector base*	*17*	*Gaiter - 2 off*
3	*Fork leg complete (right-hand) **	*18*	*Lower shroud seating - 2 off*
4	*Fork spring - 2 off*	*19*	*Fork spindle clamp - 2 off*
5	*Fork stanchion - 2 off*	*20*	*Fork top bolt - 2 off*
6	*Lower fork leg - 2 off **	*21*	*Drain plug washer - 2 off*
7	*Damper piston - 2 off*	*22*	*Circlip - 2 off*
8	*Damper valve - 2 off*	*23*	*Oil seal - 2 off*
9	*Stop ring - 2 off*	*24*	*'O' ring - 2 off*
10	*Snap ring - 2 off*	*25*	*Drain plug - 2 off*
11	*Valve stop ring - 2 off*	*26*	*Stud - 4 off*
12	*Piston stop ring - 2 off*	*27*	*Nut - 4 off*
13	*Fork bush - 2 off*	*28*	*Plain washer - 4 off*
14	*Upper shroud seating - 2 off*	*29*	*Spring washer - 4 off*
15	*Upper shroud - 2 off (handed)*	***	*Left hand identical, but handed*

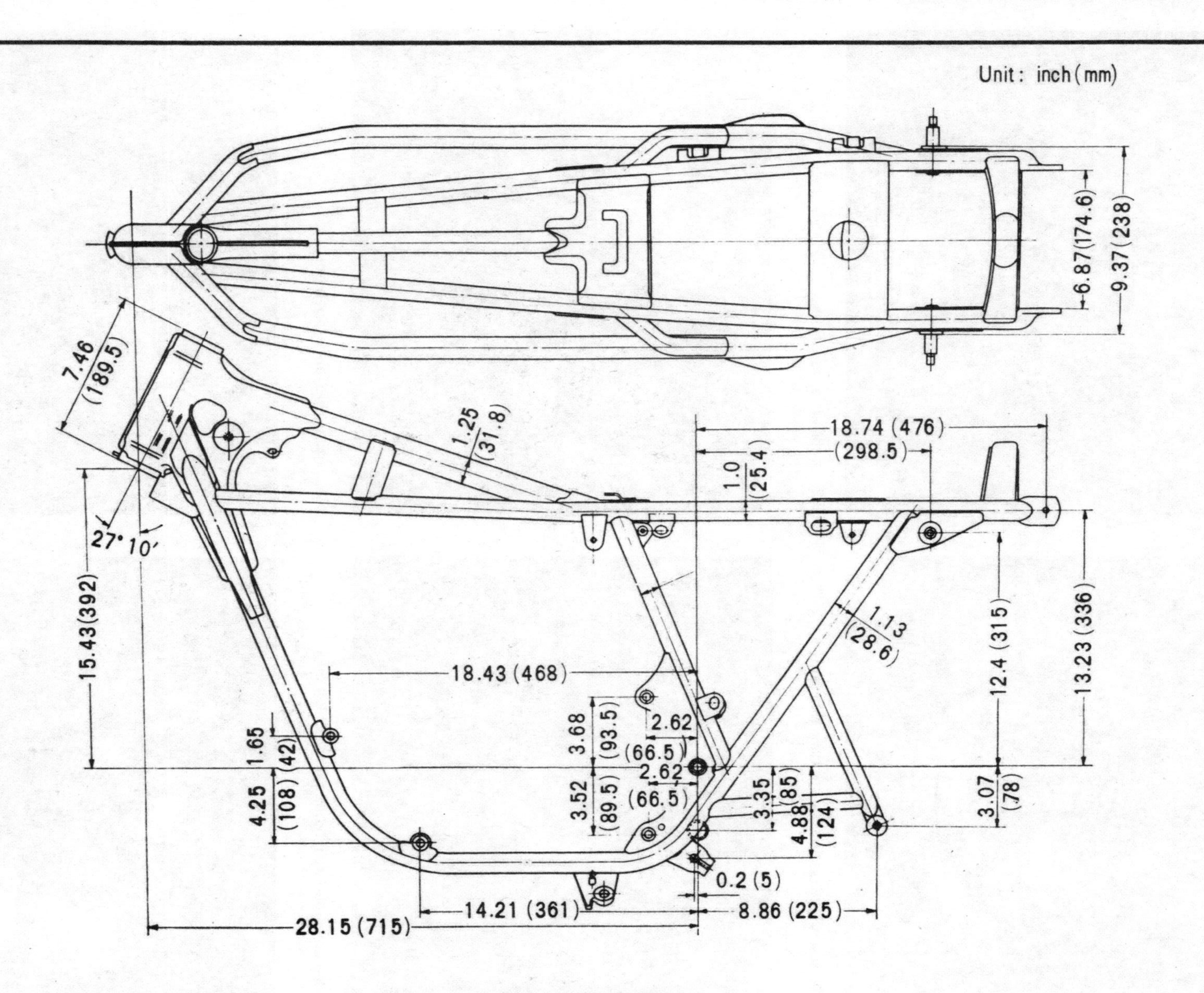

Fig. 4.2. Honda CB750 Frame dimensions

3.2b. Oil seal will be displaced at the same time and ...

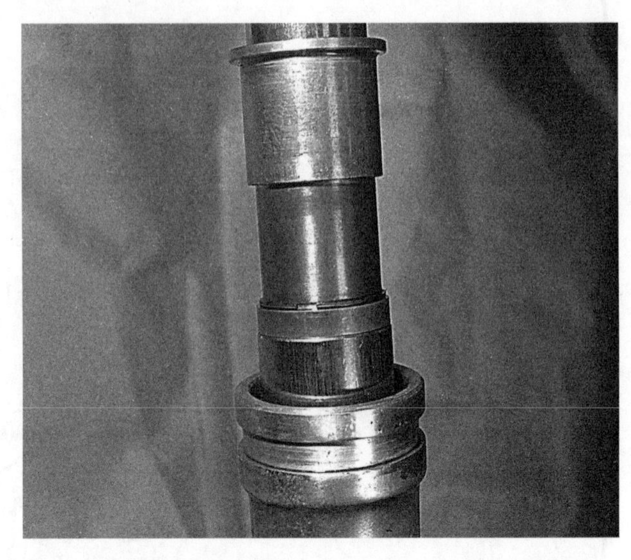

3.2c. ... the upper fork bush

3.2d. Remove circlip from bottom of stanchion tube ...

3.2e. ... to release lower fork bush

3.3. Unscrew top bolt to withdraw fork spring

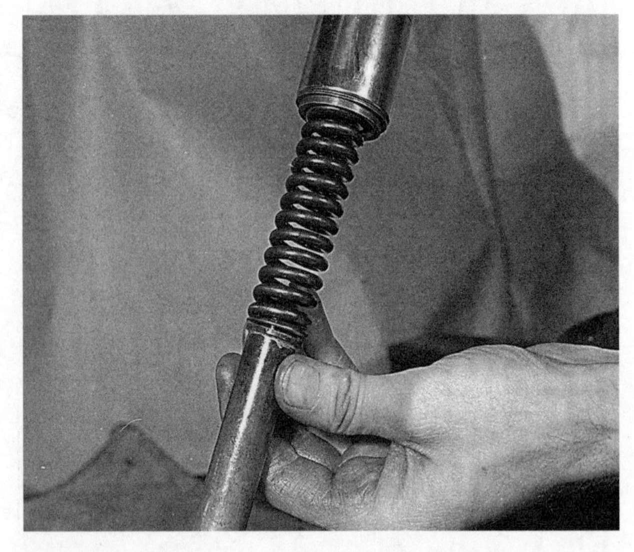

3.3a. Fork spring abuts on damper tube assembly

3 Front forks - dismantling

1 It is advisable to dismantle each fork leg separately, using an identical procedure. There is less chance of unwittingly exchanging parts if this approach is adopted. Commence by draining the fork legs; there is a drain plug in each lower leg, above and to the rear of the wheel spindle housing. Remove the chromium plated bolt at the top of the fork leg.

2 Clamp the lower fork leg in a vice fitted with soft clamps and unscrew the socket screw recessed into the housing which carries the wheel spindle. Using a pair of pointed nose pliers, remove the circlip within the top of the lower fork leg. If the fork stanchion tube is now gripped firmly and pulled sharply outward, it will come away from the lower fork leg complete with the oil seal and the upper and lower fork bushes. The oil seal and upper fork bush will pull off the stanchion in an upward direction; the lower bush is released downward after the retaining circlip has been removed.

3 The fork spring is located within the stanchion and abuts on the end of the damper tube. It can be pulled out without difficulty when the top bolt is unscrewed It is unlikely that the damper valve will need to be disturbed; it is retained in position by two circlips which act as limit stops.

4 Note that several changes in fork design have been made since the CB 750 was first manufactured. The broad dismantling procedure, as described, applies throughout.

4 Steering head bearings - examination and renovation

1 Before commencing reassembly of the forks, examine the steering head races. The ball bearing tracks of the respective cup and cone bearings should be polished and free from indentations and cracks. If signs of wear or damage are evident, the cups and cones must be renewed. They are a tight push fit and should be drifted out of position.

2 Ball bearings are relatively cheap. If the originals are marked or discoloured, they should be renewed. To hold the steel balls in position dueing re-attachment of the forks, pack the bearings with grease. Note that each race should contain a total of nineteen ¼ inch diameter ball bearings. Although space will be left to include one extra ball, it is necessary to prevent the bearings from skidding on each other and accelerating the rate of wear.

5 Front forks - examination and renovation

1 The parts most liable to wear over an extended period of service are the fork bushes which fit over the stanchions and are easily renewed. Worn fork bushes cause judder when the front brake is applied and the increased amount of play can be detected by pulling and pushing on the handlebars when the front brake is applied fully. Wear of the sliding damper piston is less common, but this too is easily renewed if the damping action deteriorates.

2 It is advisable to renew the oil seals when the forks are dismantled, even if they appear to be in good condition. This will save a further fork stripdown at a later date, should oil leakage occur. Check the condition of the rubber gaiters at the same time, since they will admit water and road grit if cracks or splits are allowed to develop. Rusty fork stanchions often create problems when dismantling is necessary because the oil seals and fork bushes cannot be extracted until the surface on which they slide has been cleaned up.

3 It is rarely possible to straighten forks which have been badly damaged in an accident, especially if the correct jigs are not available. It is always best to err on the side of safety and fit new ones, especially since there is no easy means of detecting whether the forks have been overstressed and the metal fatigued. Fork stanchions can be checked for straightness by stripping them and rolling them on a flat surface. Any misalignment will immediately be obvious.

4 The fork springs will compress after a lengthy period of service and will need to be renewed if fork action becomes spongy. The free length of new springs is 19.075 inch (484.5 mm) and the serviceable limit is reached at 18.11 inch (460 mm).

5 Fork damping is governed by the viscosity of the oil in the fork legs, normally SAE 10W-30, and by the action of the damper piston. The outside diameter of the piston is 1.552 - 1.553 inch (39.50 - 39.45 mm), but if this decreases to 1.5591 inch (39.4 mm) renewal is necessary. Each fork leg holds 220 - 230 cc (7.0 - 7.3 fluid ounces) of oil.

6.2. Do not omit to refill with correct quantity of damping oil

6.2a. Lower right-hand pinch bolt holds speedometer cable clamp

6 Front forks - replacement

1 Replace the front forks by following in reverse the dismantling procedures described in Sections 2 and 3 of this Chapter. Before fully tightening the front wheel spindle clamps and the fork yoke pinch bolts, bounce the forks several times to ensure they work freely and are clamped in their original settings. Complete the final tightening from the wheel spindle clamps upward.

2 Do not forget to add the recommended quantity of fork damping oil to each leg before the bolts in the top of each fork leg are replaced. Check that the drain plugs have been re-inserted and tightened before the oil is added.

3 If the fork stanchions prove difficult to re-locate through the fork yokes, make sure their outer surfaces are clean and polished so that they will slide more easily. It is often advantageous to use a screwdriver blade to open up the clamps as the stanchions are pushed upward into position.

4 Before the machine is used on the road, check the adjustment of the steering head bearings. If they are too slack, judder will occur. There should be no detectable play in the head races when the handlebars are pulled and pushed, with the front brake applied hard.
5 Overtight head races are equally undesirable. It is possible to unwittingly apply a loading of several tons on the head bearings by overtightening, even though the handlebars appear to turn quite freely. Overtight bearings will cause the machine to roll at low speeds and give generally imprecise handling with a tendency to weave. Adjustment is correct if there is no perceptible play in the bearings and the handlebars will swing to full lock in either direction, when the machine is on the centre stand with the front wheel clear of the ground. Only a slight tap should cause the handlebars to swing.

7 Steering head lock

1 The steering head lock is attached to the underside of the lower yoke of the forks by a single screw and washer. When in a locked position, a tongue extends from the body of the lock when the handlebars are on full lock in either direction and abuts against a plate welded to the base of the steering head. In consequence, the handlebars cannot be straightened until the lock is released.
2 If the lock malfunctions, it must be renewed. A repair is impracticable. When the lock is changed the key must be changed too, to match the new lock. No maintenance is necessary; when not in use the lock is protected by a spring-loaded cover plate.

8 Frame - examination and renovation

1 The frame is unlikely to require attention unless accident damage has occurred. In some cases, replacement of the frame is the only satisfactory course of action if it is badly out of alignment. Only a few frame repair specialists have the jigs and mandrels necessary for resetting the frame to the required standard of accuracy and even then there is no easy means of assessing to what extent the frame may have been overstressed.
2 After the machine has covered a considerable mileage, it is advisable to examine the frame closely for signs of cracking or splitting at the welded joints. Rust can also cause weakness at these joints. Minor damage can be repaired by welding or brazing, depending on the extent and nature of the damage.
3 Remember that a frame which is out of alignment will cause handling problems and may even promote 'speed wobbles'. If misalignment is suspected, as the result of an accident, it will be necessary to strip the machine completely so that the frame can be checked and, if necessary, renewed.

9 Swinging arm rear fork - dismantling, examination and renovation

1 The rear fork of the frame assembly pivots on a detachable bush within each end of the fork crossmember and a pivot shaft which itself is surrounded by a long, detachable bush. The pivot shaft passes through frame lugs on each side of the engine unit and the centre of the long detachable bush, so that the inner and outer bushes form the bearing surfaces. It is quite easy to renovate the swinging arm when wear necessitates attention.
2 To remove the swinging arm fork, first position the machine on the centre stand, then detach the plastics chainguard which is retained by two bolts and a metal clip. Detach the final drive chain from the rear wheel sprocket by removing the spring link (early models). Detach the fork ends to release rear wheel spindle from frame.
3 Detach the rear brake torque arm from the brake plate on the right hand side of the machine by removing the spring clip, then the retaining nut, bolt and washer. Remove the rear brake rod by unscrewing the adjuster from the end and pulling the rod from

9.4. Detach rear suspension units at lower end only

9.7. Swinging arm fork will pull from frame after pivot shaft is withdrawn

9.7a. Dust covers protect fork bushes

9.11. Brake arm must be IN FRONT of swinging arm after re-assembly

9.11a. End of rear chainguard is retained by a clip

the trunnion within the brake operating arm. Take care not to lose the trunnion which is now free. Remove the split pin from the left hand end of the rear wheel spindle and unscrew the castellated nut. Before the wheel spindle is withdrawn, slacken the chain adjusters and push the wheel forward so that the chain can be lifted off the sprocket teeth (late models only). Do not misplace the distance piece between the brake plate and the inside of the fork end which will fall clear.

4 Remove the brake plate complete with the rear brake assembly; this will lift away from the wheel. Then remove the rear wheel complete with sprocket by tilting it so that there is sufficient clearance to ease it from under the mudguard and away from the machine.

5 Remove both rear suspension units from the swinging arm fork. Each is bolted at the bottom end to the swinging arm fork, through a rubber bush. The top mounting takes the form of a stud projecting from the subframe which passes through the upper rubber bush mounting of the unit. The unit is secured by an acorn nut which need be slackened only.

6 Remove the rear brake torque arm which is retained by a single bolt at the forward end, then remove the locknut from the end of the fork pivot shaft and withdraw the shaft itself. The swinging arm fork is now free to be withdrawn from the frame.

7 Removal is made easier if the fork is first raised and then pushed forward so that the dust covers can be removed from each end of the pivot housing. They pull off quite easily. Then draw the fork backwards again and tilt the left hand end upward. It should then clear the frame without difficulty.

8 The two pivot bushes should press out of the ends of the fork crossmember with ease. The inner bush which surrounds the pivot shaft will also be released at the same time. Wash the bearings and the pivot shaft with a petrol/paraffin mix, then check the amount of play between them. If the clearance exceeds 0.5 mm (0.020 in) the bearings and the pivot shaft should be renewed as a set. If renewal is not necessary, check the pivot shaft for straightness. If it is bent, it must be renewed.

9 Reassemble the swinging arm fork by reversing the dismantling procedure. Grease the pivot shaft and bearings liberally prior to reassembly and check that the seals within the dust covers are in good order.

10 Worn swinging arm pivot bearings will give imprecise handling with a tendency for the rear end of the machine to twitch or hop. The play can be detected by placing the machine on its centre stand and with the rear wheel clear of the ground, pulling and pushing on the fork ends in a horizontal direction. Any play will be greatly magnified by the leverage effect.

11 If for any reason the rear brake pedal has been removed, check to make sure the arm on the pivot to which the rear brake rod is attached is positioned IN FRONT of the swinging arm crossmember (nearest to the engine). It is very difficult to re-locate without first removing the swinging arm assembly or the return spring, the latter being quite difficult to re-tension in situ.

10 Rear suspension units - examination

1 The rear suspension units fitted to the earlier models were of the hydraulically-damped three-position type, adjustable to give three different spring settings. Either a 'C' spanner or a metal rod is inserted into the peg holes immediately above the adjusting notches, to turn the adjusters. Rotate clockwise to increase the spring tension and stiffen up the suspension. The recommended settings are as follows:

Position 1 (least tension)	Normal running, without a pillion passenger
Position 2 (middle position)	High speed touring
Position 3 (most tension)	Very high speed or with pillion passenger and/or heavy loads

2 Later models used nitrogen-filled units with a similar range of adjustment and these in turn have now been replaced by units of the more conventional hydraulically-damped type which have five adjustable settings.

3 There is no means of draining the units or topping up, because the dampers are built as a sealed unit. If the damping fails or if the units commence to leak, the complete damper assembly must be renewed.

4 The compression springs can be removed by detaching the damper units from the machine and holding each one upright on the workbench whilst the outer top shroud is pressed downward in opposition to the spring pressure. This will permit a second person to remove the split collets from the top of the shroud so that the shroud and spring can be withdrawn over the upper end of the unit. The spring should have a free length of 8.74 inch (222 mm) and should be renewed if it has compressed to 8.504 inch (216 mm) or beyond.

5 In the interests of good road holding, it is essential that both units are set to the same load setting and that when replacements are made, both units are treated in an identical manner so that they stay matched.

11 Centre stand - examination

1 The centre stand pivots on a hollow tube clamped between two split lugs which extend from the rearmost crossmember of the frame. The clamps are held tight by a pinch bolt arrangement

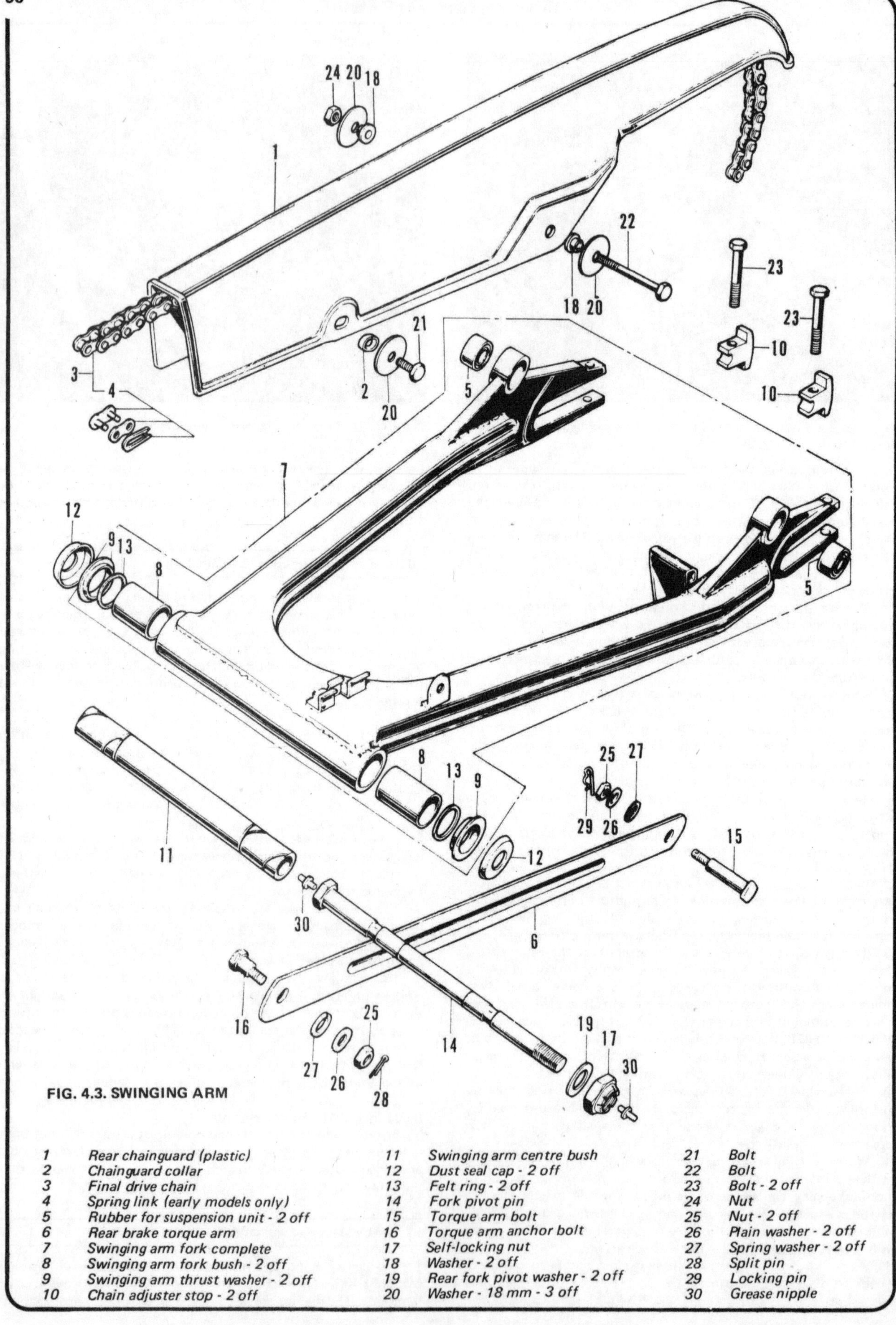

FIG. 4.3. SWINGING ARM

1	*Rear chainguard (plastic)*	11	*Swinging arm centre bush*	21	*Bolt*
2	*Chainguard collar*	12	*Dust seal cap - 2 off*	22	*Bolt*
3	*Final drive chain*	13	*Felt ring - 2 off*	23	*Bolt - 2 off*
4	*Spring link (early models only)*	14	*Fork pivot pin*	24	*Nut*
5	*Rubber for suspension unit - 2 off*	15	*Torque arm bolt*	25	*Nut - 2 off*
6	*Rear brake torque arm*	16	*Torque arm anchor bolt*	26	*Plain washer - 2 off*
7	*Swinging arm fork complete*	17	*Self-locking nut*	27	*Spring washer - 2 off*
8	*Swinging arm fork bush - 2 off*	18	*Washer - 2 off*	28	*Split pin*
9	*Swinging arm thrust washer - 2 off*	19	*Rear fork pivot washer - 2 off*	29	*Locking pin*
10	*Chain adjuster stop - 2 off*	20	*Washer - 18 mm - 3 off*	30	*Grease nipple*

and periodically these bolts should be checked to ensure that they are tight. An extension spring is used to keep the stand in the fully retracted position whilst the machine is in motion.
2 Check that the return spring is in good condition and that the stand pivot is neither tight nor sloppy. A broken or weak spring may cause the stand to fall whilst the machine is being ridden and catch in some obstacle, unseating the rider.

12 Prop stand - examination

1 The prop stand bolts to a lug attached to the rear of the left hand lower frame tube. An extension spring ensures that the stand is retracted when the weight of the machine is taken off the stand.
2 Check that the pivot bolt is secure and that the extension spring is in good condition and not over-stretched. An accident is almost inevitable if the stand extends whilst the machine is on the move.

13 Footrests - examination and renovation

1 The footrests bolt to a rod which passes through the frame gussets on either side of the machine, at the point where the lower frame tubes commence to curve upward. The footrests pivot upwards and are spring-loaded to keep them in the normal horizontal position. If an obstacle is struck, they will fold upwards, obviating the risk of injury to the rider's foot or damage to the frame.
2 If the footrests are damaged in an accident, it is possible to dismantle the assembly into its component parts by detaching each footrest from the rod and separating the folding rubber from the main support on which it pivots by withdrawing the split pin and releasing the clevis pin through the pivot. It is preferable to renew the damaged parts, but if necessary they can be bent straight by clamping them in a vice and heating to a dull red with a blow lamp whilst the appropriate pressure is applied. Never attempt to bend the footrests straight whilst they are still attached to the frame.
3 If heat is applied to the main footrest support during any straightening operation, it follows that the footrest rubber must be removed temporarily to prevent damage from heat conduction.

14 Rear brake pedal - examination and renovation

1 The rear brake pedal is attached to a splined pivot which passes through a tube welded to the lower right hand frame tube, immediately below the swinging arm pivot. A short arm welded to the other end of the pivot forms the point of attachment for the rod which operates the rear brake. The pivot is fitted with a strong return spring coiled around its boss, to provide the brake pedal with positive action.
2 If the brake pedal is bent or twisted, it can be drawn off the splined pivot and straightened by adopting the same technique as recommended for bent footrests. It is held in position by a single clamp bolt. Mark the position of the pedal in relation to the pivot splines before removal, so that it is replaced in the same position.
3 Check the return spring periodically and renew it if it shows signs of weakening.

15 Dualseat - removal and replacement

1 The seat is attached to small lugs on the right hand side of the subframe and pivots on two long clevis pins which pass through these lugs. It opens from the left, after a safety catch has been released and on later models locks in position to safeguard the battery and safety helmets attached to hooks covered by the seat.
2 Although it is seldom necessary to detach the seat, it can be lifted away as a complete unit if the split pins through the ends of the clevis pins are withdrawn, and then the clevis pins themselves.

16 Speedometer and tachometer heads - removal and replacement

1 The speedometer and tachometer heads are freed quite readily by slackening the chromium-plated clamp around both instruments, so that they can be lifted upward. Detach the drive cables at the screwed couplings and unscrew the two crosshead screws which retain the underplate below each instrument, so that the bulb holders can be withdrawn from the base.
2 Apart from defects in either the drive or drive cables, a speedometer or tachometer which malfunctions is difficult to repair. Fit a replacement or alternatively entrust the repair to a competent instrument repair specialist.
3 Remember that a speedometer in correct working order is a statutory requirement in the UK. Apart from this legal necessity, reference to the odometer readings is the most satisfactory means of keeping pace with the maintenance schedules.

17 Speedometer and tachometer drive cables - examination and maintenance

1 It is advisable to detach the drive cable(s) from time to time in order to check whether they are lubricated adequately, and whether the outer coverings are damaged or compressed at any point along their run. Jerky or sluggish movements can often be traced to a damaged drive cable.
2 For greasing, withdraw the inner cable. After removing all the old grease, clean with a petrol-soaked rag and examine the cable for broken strands or other damage.
3 Regrease the cable with high melting point grease, taking care not to grease the last six inches at the point where the cable enters the instrument head. If this precaution is not observed, grease will work into the head and immobilise the instrument movement.
4 If any instrument head stops working suspect a broken drive cable unless the odometer readings continue. Inspection will show whether the inner cable has broken; if so, the inner cable alone can be replaced and re-inserted in the outer casing, after greasing. Never fit a new inner cable alone if the outer covering is damaged or compressed at any point along its run.

18 Speedometer and tachometer drives - location and examination

1 The speedometer drive gearbox forms part of the front brake plate assembly and is driven internally from the wheel hub. The tongued drive pinion engages with slots formed on the left hand end of the wheel hub. The drive rarely gives trouble; it is pre-packed with grease and should be relubricated when the front wheel bearings receive attention.
2 The tachometer drive is taken from the cylinder head cover, between number three and four cylinders. The drive is taken from the overhead camshaft by means of skew-cut pinions, and then by a flexible cable to the tachometer head. It is unlikely that the drive will give trouble during the normal service life of the machine, especially since it is fully enclosed and effectively lubricated.

19 Cleaning the machine

1 After removing all surface dirt with a rag or sponge which is washed frequently in clean water, the machine should be allowed to dry thoroughly. Application of car polish or wax to the cycle parts will give a good finish, particularly if the machine receives this attention at regular intervals.
2 The plated parts should require only a wipe with a damp rag,

but if they are badly corroded, as may occur during the winter when the roads are salted, it is permissible to use one of the proprietary chrome cleaners. These often have an oily base which will help to prevent corrosion from recurring.
3 If the engine parts are particularly oily, use a cleaning compound such as Gunk or Jizer. Apply the compound whilst the parts are dry and work it in with a brush so that it has an opportunity to penetrate and soak into the film of oil and grease. Finish off by washing down liberally, taking care that water does not enter the carburettors, air cleaners or the electrics. If desired, the now clean aluminium alloy parts can be enhanced still further when they are dry by using a special polish such as Solvol Autosol. This will restore the full lustre.
4 If possible, the machine should be wiped down immediately after it has been used in the wet, so that it is not garaged under damp conditions which will promote rusting. Make sure that the chain is wiped and re-oiled, to prevent water from entering the rollers and causing harshness with an accompanying rapid rate of wear. Remember there is less chance of water entering the control cables and causing stiffness if they are lubricated regularly as described in the Routine Maintenance Section.

20 Sidecar alignment

1 Although the Honda CB 750 has no lugs for the attachment of a sidecar, it is possible to fit one if the appropriate fittings are available.
2 Good handling characteristics of the outfit will depend on the accuracy with which the sidecar is aligned. Provided the toe-in and lean-out are within prescribed limits, good handling characteristics should result, leaving scope for other minor adjustments about which opinions vary quite widely.
3 To set the toe-in, check that the front and rear wheels of the motor cycle are correctly in line and adjust the sidecar fittings so that the sidecar wheel is approximately parallel to a line drawn between the front and rear wheels of the machine. Re-adjust the fittings so that the sidecar wheel has a slight toe-in toward the front wheel of the motor cycle, as shown in Fig. 4.4. When the amount of toe-in is correct, the distance 'B' should be from 3/8 inch to 3/4 inch less than the distance at 'A'.
4 Lean-out is checked by attaching a plumb line to the handlebars and measuring the distance between 'C' and 'D' as shown in Fig. 5.4. Lean out is correct when the distance 'C' is approximately 1 inch greater than at 'D'.

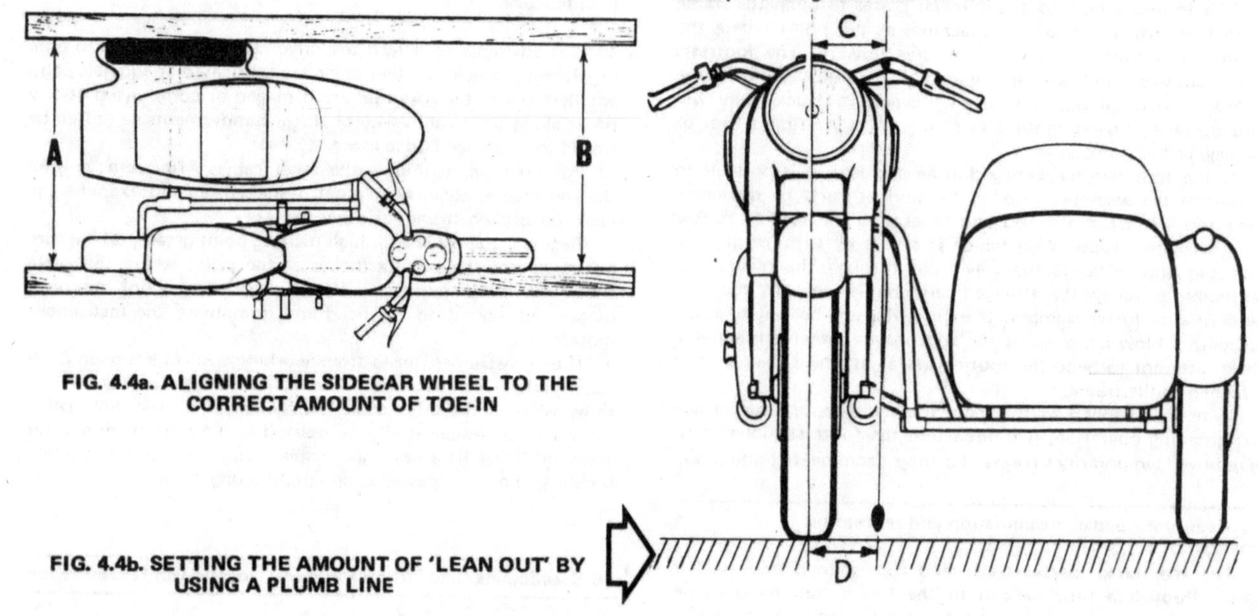

FIG. 4.4a. ALIGNING THE SIDECAR WHEEL TO THE CORRECT AMOUNT OF TOE-IN

FIG. 4.4b. SETTING THE AMOUNT OF 'LEAN OUT' BY USING A PLUMB LINE

21 Fault diagnosis

Symptom	Reason/s	Remedy
Machine veers to left or right with hands off handlebars	Incorrect wheel alignment Bent forks Twisted frame	Check and re-align. Check and renew. Check and renew.
Machine rolls at low speeds	Overtight steering head bearings	Slacken and re-test.
Machine judders when front brake is applied	Slack steering head bearings	Tighten until all play is taken up.
Machine pitches badly on uneven surfaces	Ineffective fork dampers Ineffective rear suspension units	Check oil content. Check damping action.
Fork action stiff	Fork legs out of alignment (twisted yokes)	Slacken yoke clamps, front wheel spindle and fork top bolts. Pump forks several times, then tighten from bottom upward.
Machine wanders. Steering imprecise, rear wheel tends to hop	Worn swinging arm pivot	Dismantle and renew bushes and pivot shaft.

Chapter 5 Wheels Brakes and Tyres

Contents

Specifications

Tyres

Front	3.25 x 19 in
Rear	4.00 x 18 in

Tyre pressures

Front	28 psi	32 psi *
Rear	28 psi	34 psi *

* Higher pressures for sustained high speeds (over 110 mph)

Brakes

Front	Hydraulic disc brake
Rear	7 inch diameter single leading shoe drum brake

1 General description

The Honda CB 750 has a 19 inch diameter front wheel and an 18 inch diameter rear wheel. The front tyre is of 3.25 inch section and normally has a block tread pattern. The rear tyre is of 4.00 section and has a similar tread formation. The front wheel is fitted with an hydraulically-operated disc brake and is quickly detachable. The rear wheel has a conventional single leading shoe drum brake of approximately 7 inch diameter. It is not of the quickly detachable type, necessitating removal of the final drive chain before the wheel can be detached from the frame.

2 Front wheel - examination and renovation

1 Place the machine on the centre stand so that the front wheel is raised clear of the ground. Spin the wheel and check the rim alignment. Small irregularities can be corrected by tightening the spokes in the affected area although a certain amount of experience is necessary to prevent over-correction. Any flats in the wheel rim will be evident at the same time. These are more difficult to remove and in most cases it will be necessary to have the wheel rebuilt on a new rim. Apart from the effect on stability, a flat will expose the tyre bead and walls to greater risk of damage if the machine is run with a deformed wheel.
2 Check for loose and broken spokes. Tapping the spokes is the best guide to tension. A loose spoke will produce a quite different sound and should be tightened by turning the nipple in an anticlockwise direction. Always check for run out by spinning the wheel again. If the spokes have to be tightened by an excessive amount, it is advisable to remove the tyre and tube as detailed in Section 15 of this Chapter. This will enable the protruding ends of the spokes to be ground off, thus preventing them from chafing the inner tube and causing punctures.

3 Front wheel disc brake - examination and renovation

1 Check the front brake master cylinder, hose and caliper unit for signs of fluid leakage. Pay particular attention to the condition of the hose, which should be renewed without question if there are signs of cracking, splitting or other exterior damage.
2 Check the level of hydraulic fluid by removing the cap on the brake fluid reservoir, diaphragm plate and diaphragm. This is one of the regular maintenance tasks, which should never be neglected. If the fluid is below the level mark, brake fluid of the correct grade must be added. NEVER USE ENGINE OIL or anything other than the recommended fluid. Other fluids have unsatisfactory characteristics and will rapidly destroy the seals.
3 The brake pads should be inspected for wear. Each has a red tab which marks the limit of the friction material (late models only). When this limit has been reached, BOTH pads must be renewed, even if only one has reached the limit line. Check by applying the brake so that the pads engage with the disc. They will lift out of the caliper unit when the front wheel is removed and the caliper unit is separated. On early models replace the pads when the clearance between the front of the caliper and the disc exceeds 0.06 inch (1.5 mm). See Section 4 of this Chapter.
4 If brake action becomes spongy, or if any part of the hydraulic system is dismantled (such as when the hose is

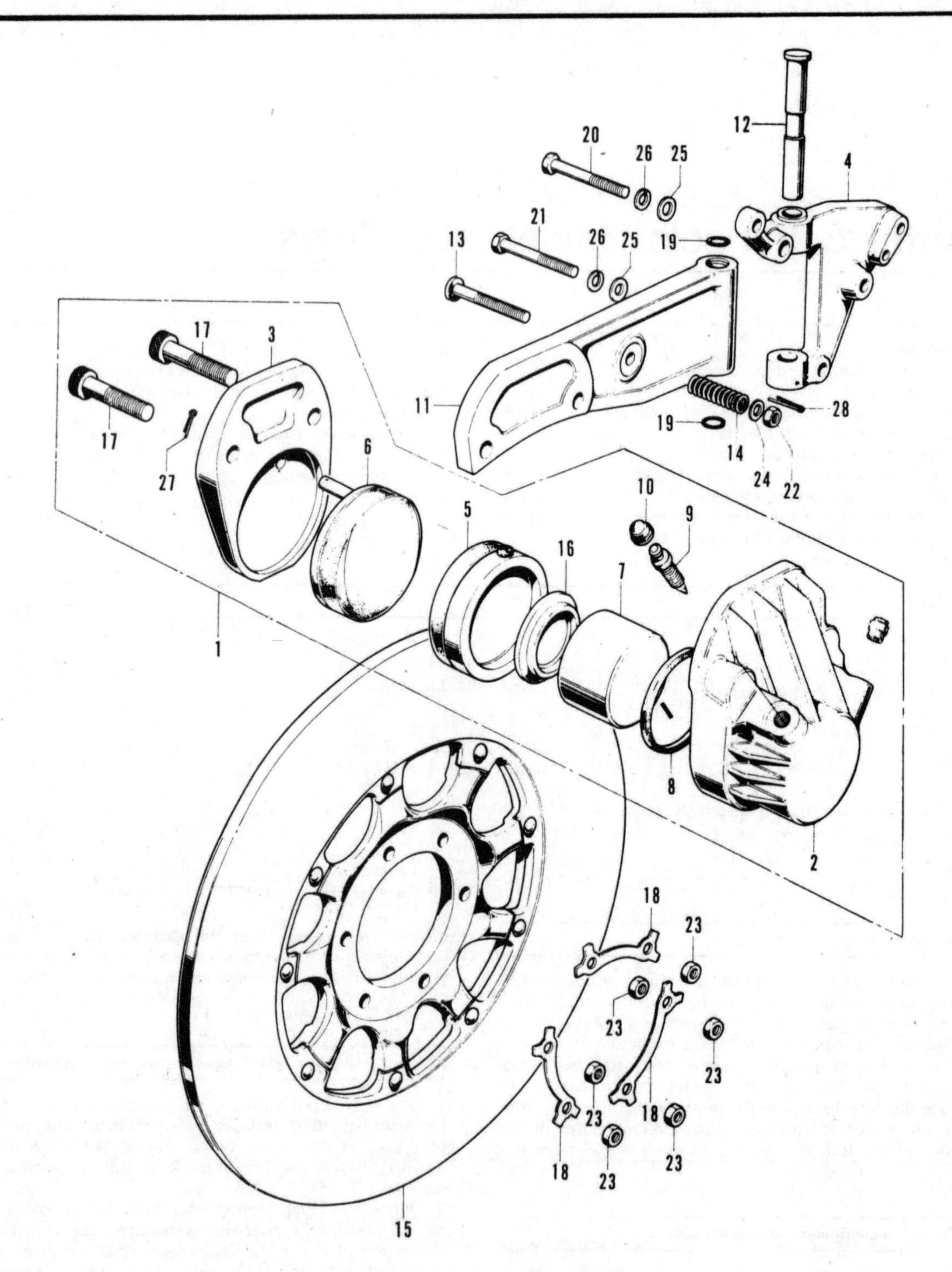

FIG. 5.1. FRONT DISC BRAKE

1	*Caliper assembly complete*	*11*	*Caliper holder*	*21*	*Bolt*
2	*Left-hand caliper*	*12*	*Caliper holder pivot pin*	*22*	*Nut*
3	*Right-hand caliper*	*13*	*Caliper adjusting bolt*	*23*	*Nut - 6 off*
4	*Caliper holder pivot*	*14*	*Caliper adjusting spring*	*24*	*Plain washer*
5	*Left-hand pad*	*15*	*Front brake disc*	*25*	*Plain washer - 3 off*
6	*Right-hand pad*	*16*	*Pad seating*	*26*	*Spring washer*
7	*Piston*	*17*	*Caliper socket screws - 2 off*	*27*	*Split pin*
8	*Seal*	*18*	*Tongued washer - 3 off*	*28*	*Split pin*
9	*Bleeder nipple*	*19*	*'O' ring - 2 off*		
10	*Cap for bleeder*	*20*	*Bolt - 2 off*		

replaced) it is necessary to bleed the system in order to remove all traces of air. The following procedure should be followed:

5 Attach a tube to the bleed valve at the top of the caliper unit, after removing the dust cap. It is preferable to use a transparent plastics tube, so that the presence of air bubbles is seen more readily.

6 The far end of the tube should rest in a small bottle so that it is submerged in hydraulic fluid. This is essential, to prevent air from passing back into the system. In consequence, the end of the tube must remain submerged at all times.

7 Check that the reservoir on the handlebars is full of fluid and replace the cap to keep the fluid clean.

8 If spongy brake action necessitates the bleeding operation, squeeze and release the brake lever several times in rapid succession, to allow the pressure in the system to build up. Then open the bleed valve by unscrewing it one complete turn whilst maintaining pressure on the lever. This is a two-person operation. Squeeze the lever fully until it meets the handlebar, then close the bleed valve. If parts of the system have been replaced, the bleed valve can be opened from the beginning and the brake lever worked until fluid issues from the bleed tube. Note that it may be necessary to top up the reservoir during this operation; if it empties, air will enter the system and the whole operation will have to be repeated.

9 Repeat operation 8 until bubbles disappear from the bleed tube. Close the bleed valve fully, remove the bleed tube and replace the dust cap.

10 Check the level in the reservoir and top up if necessary. Never use the fluid which has drained into the bottle at the end of the bleed tube because this contains air bubbles which will re-introduce air into the system. It must stand for 24 hours before it can be re-used.

11 Refit the diaphragm and diaphragm plate and tighten the reservoir cap securely.

12 Do not spill hydraulic fluid on the cycle parts. It is a very effective paint stripper! On early models the plastics 'glasses' in the speedometer and tachometer heads will be obscured badly if hydraulic fluid is spilt on them. Later models have glass, which is unaffected.

4 Replacing brake pads and overhauling the caliper unit

1 Remove the front wheel by following the procedure described in Chapter 4, Section 2.7. Before the friction pads can be detached, it will be necessary to separate the two parts of the caliper unit by unscrewing the two large diameter socket head screws on the inside facing finned cover. These screws are extremely tight. When the caliper unit is separated, the left hand pad can be lifted out. Note how an extending peg at the rear engages with a hole in the unit, as a means of location. The right hand pad has a similar peg, but before this pad can be withdrawn from the caliper, it is first necessary to remove the split pin which passes through the peg on the outside of the unit.

2 Inspect the friction pads closely and renew them both if the limit level of wear is approached, as described in paragraph 3 of the preceding Section. If there is any doubt whatsoever about their condition, they should be renewed as a pair.

3 Clean the recesses into which the pads fit and the exposed end of the piston which actuates the left hand pad. Use only a small, soft brush and NOT solvent or a wire brush. Smear the piston face and the brake pad recesses with hydraulic fluid, to act as a lubricant. Only sparing lubrication is required.

4 Remove the reservoir cap, diaphragm plate and diaphragm to check whether the level of fluid rises as the pistons are pushed back into the recesses. It may be necessary to syphon some fluid out of the reservoir prior to this operation, to prevent over-flowing. If the piston does not move freely, the caliper must be removed from the machine and overhauled. Because damage of some kind is inevitable to cause piston seizure, it is best to entrust the repair or replacement of the unit to a Honda agent.

5 To check or re-set the brake caliper adjustment, reassemble the unit and refit the front wheel. Arrange the machine so

3.5. End of bleed tube must remain submerged in brake fluid

4.1. Two socket screws are very tight

4.1a. Left-hand brake pad has a locating peg

that the front wheel is clear of the ground. Loosen the locknut on the caliper adjuster and turn the adjuster until the friction pad contacts the disc to provide some resistance to the wheel's motion. Slacken off slightly until the wheel again revolves quite freely, then turn back in the SAME direction (slackening off) a further 1/8 to 1/4 turn before tightening the locknut. Caliper adjustment is now correct.

5 Removing and replacing the brake disc

1 It is unlikely that the disc will require attention unless it is badly scored and braking efficiency is reduced. To remove the disc, first detach the front wheel from the forks as described in Chapter 4, Section 2.7. The disc is bolted to the left hand side of the wheel by six long bolts, the nuts of which are secured in pairs by a common tab washer. Bend back the tab washers and remove the nuts, to free the disc.
2 Replace the disc by reversing the dismantling procedure. Make sure that all six nuts are tightened fully and that the tab washers are bent back into position.

6 Master cylinder - examination and renovation

1 The master cylinder is unlikely to give trouble unless the machine has been stored for a lengthy period or until a considerable mileage has been covered. The usual signs of trouble are leakage of hydraulic fluid and a gradual fall in the fluid reservoir content.
2 To gain full access to the master cylinder, commence the dismantling operation by attaching a bleed tube to the caliper unit bleed nipple. Open the bleed nipple one complete turn, then operate the front brake lever until all fluid is pumped out of the reservoir. Close the bleed nipple, detach the tube and store the fluid in a closed container for subsequent re-use.
3 Detach the hose and also the stop lamp switch (if fitted). Remove the handlebar lever pivot bolt and the lever itself.
4 Access is now available to the piston and the cylinder and it is possible to remove the piston assembly, together with all the relevant seals. Take note of the way in which the seals are arranged because they must be replaced in the same order. Failure to observe this necessity will result in brake failure.
5 Clean the master cylinder and piston with either hydraulic fluid or alcohol. On no account use either abrasives or other solvents such as petrol. If any signs of wear or damage are evident, renewal is necessary. It is not practicable to reclaim either the piston or the cylinder bore.
6 Soak the new seals in hydraulic fluid for about 15 minutes prior to fitting, then reassemble the parts IN EXACTLY THE SAME ORDER, using the reversal of the dismantling procedure. Lubricate with hydraulic fluid and make sure the feather edges of the various seals are not damaged.
7 Refit the assembled master cylinder unit to the handlebar, and reconnect the handlebar lever, hose, stop lamp etc. Refill the reservoir with hydraulic fluid and bleed the entire system by following the procedure detailed in Section 3.4 of this Chapter.
8 Check that the brake is working correctly before taking the machine on the road, to restore pressure and align the pads correctly. Use the brake gently for the first 50 miles or so to enable all the new parts to bed down correctly.
9 It should be emphasised that repairs to the master cylinder are best entrusted to a Honda agent, or alternatively, that the defective part should be replaced by a new unit. Dismantling and reassembly requires a certain amount of skill and it is imperative that the entire operation is carried out under cleaner than average conditions.

7 Front wheel bearings - examination and replacement

1 Place the machine on the centre stand and remove the front wheel as described in Chapter 4, Section 2.7. Detach the speedometer drive gearbox, then unscrew and remove the six

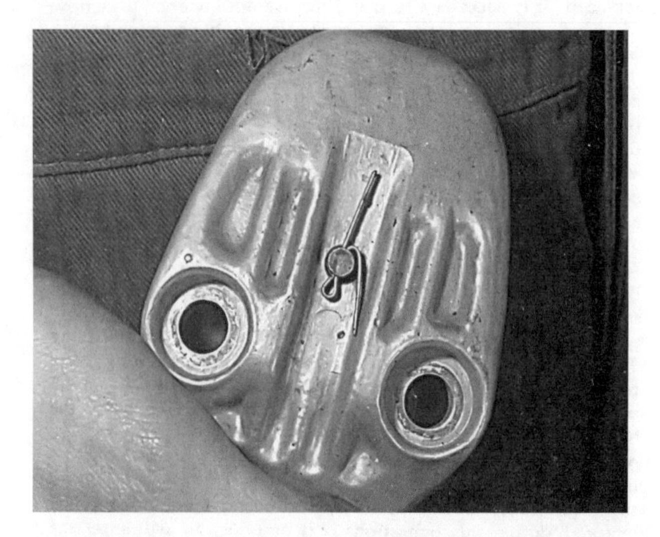

4.1b. Right-hand brake pad is secured by a split pin

4.2. Inspect pads for extent of wear

5.1. Disc is secured by six long bolts through hub

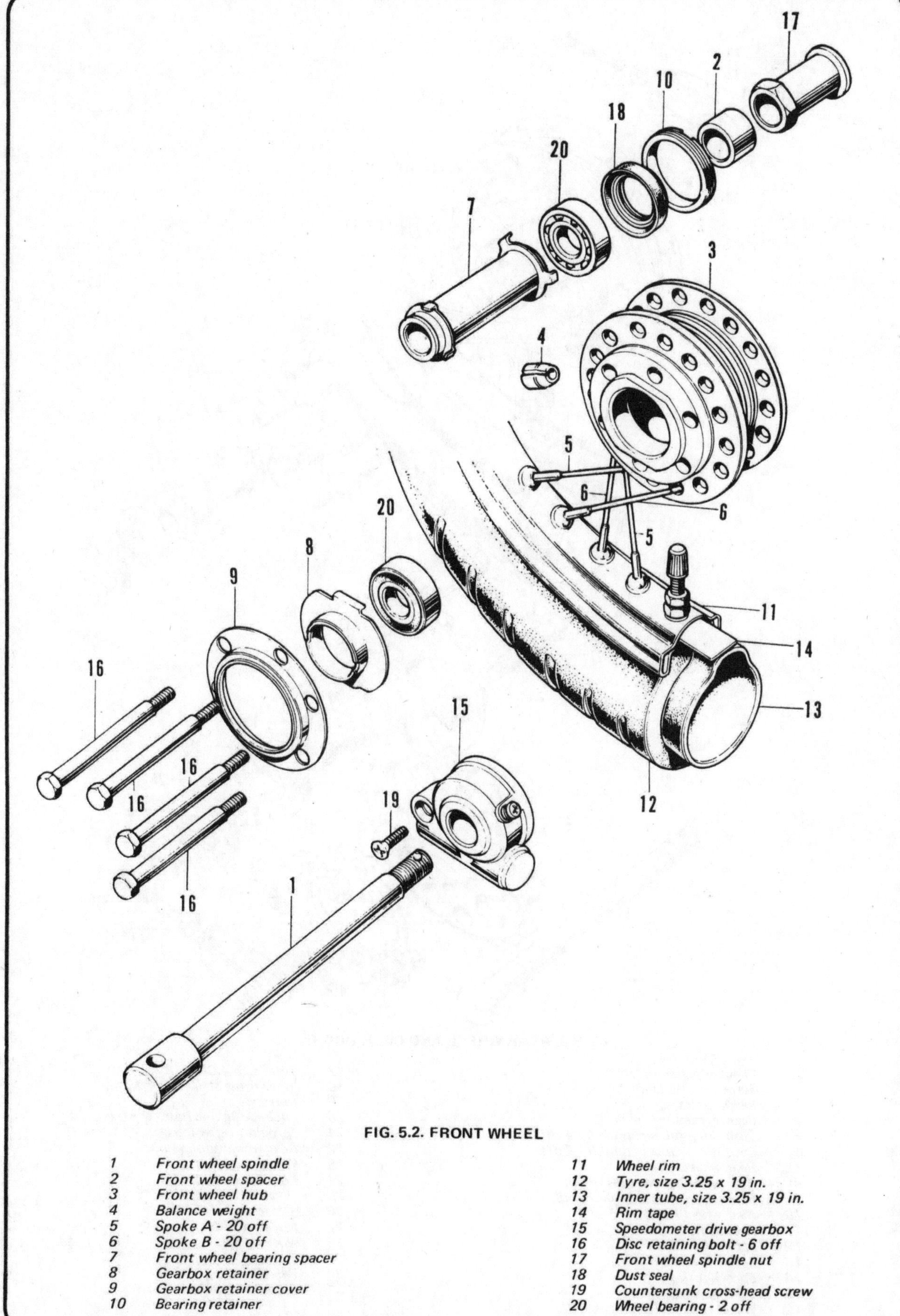

FIG. 5.2. FRONT WHEEL

1	*Front wheel spindle*	*11*	*Wheel rim*
2	*Front wheel spacer*	*12*	*Tyre, size 3.25 x 19 in.*
3	*Front wheel hub*	*13*	*Inner tube, size 3.25 x 19 in.*
4	*Balance weight*	*14*	*Rim tape*
5	*Spoke A - 20 off*	*15*	*Speedometer drive gearbox*
6	*Spoke B - 20 off*	*16*	*Disc retaining bolt - 6 off*
7	*Front wheel bearing spacer*	*17*	*Front wheel spindle nut*
8	*Gearbox retainer*	*18*	*Dust seal*
9	*Gearbox retainer cover*	*19*	*Countersunk cross-head screw*
10	*Bearing retainer*	*20*	*Wheel bearing - 2 off*

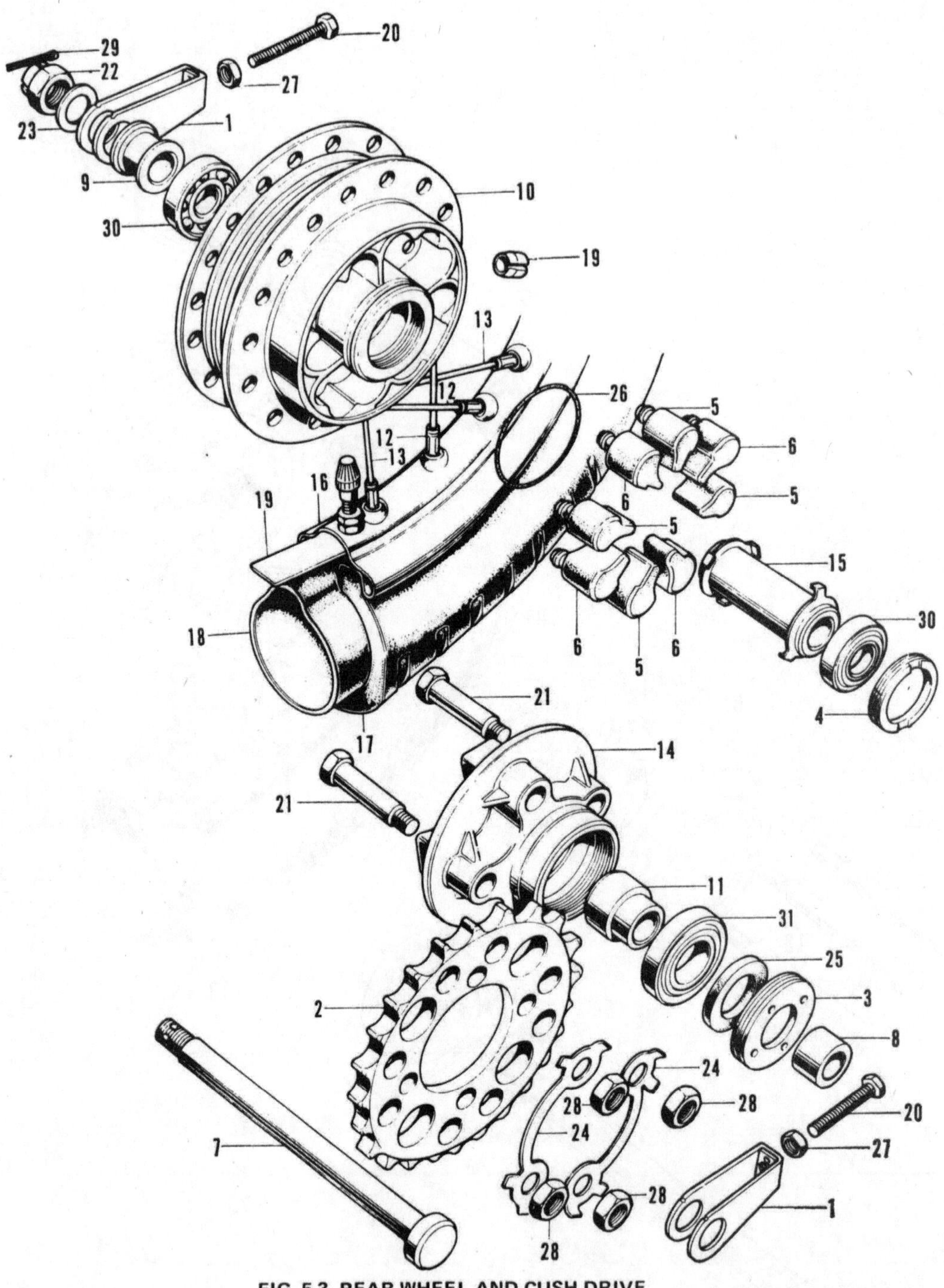

FIG. 5.3. REAR WHEEL AND CUSH DRIVE

1	*Chain adjuster - 2 off*	*17*	*Tyre, size 4.00 x 19 in.*
2	*Sprocket - 45 teeth **	*18*	*Inner tube, size 4.00 x 19 in.*
3	*Bearing retainer*	*19*	*Rim tape*
4	*Bearing retainer*	*20*	*Chain adjusting bolt - 2 off*
5	*Cush drive rubber, type A - 4 off*	*21*	*Sprocket retaining bolt - 4 off*
6	*Cush drive rubber, type **B** - 4 off*	*22*	*Rear wheel spindle nut*
7	*Rear wheel spindle*	*23*	*Washer*
8	*Rear wheel spacer (left-hand)*	*24*	*Lock washer - 2 off*
9	*Rear wheel spacer (right-hand)*	*25*	*Oil seal*
10	*Rear wheel hub*	*26*	*'O' ring*
11	*Rear spindle sleeve*	*27*	*Nut - 2 off*
12	*Spoke A - 20 off*	*28*	*Nut - 4 off*
13	*Spoke B - 20 off*	*29*	*Split pin*
14	*Sprocket flange*	*30*	*Wheel bearing - 2 off*
15	*Bearing spacer*	*31*	*Sprocket bearing*
16	*Wheel rim*		

long bolts which secure the brake disc to the hub. The nuts are retained by tab washers, which must be bent back first. When the bolts are withdrawn and the disc is removed, the flanged cover can be pulled off the right hand end of the hub. This will release the speedometer drive gearbox retainer.
2 Unscrew the bearing retainer from the left hand end of the hub and remove the dust seal behind it. The bearing can then be drifted out from the right hand side of the hub by using a two-diameter drift which will locate with the inner diameter of the distance collar behind the bearing. The collar will push the bearing out of position. When both the bearing and the collar are free, drive out the right hand bearing, using the same drift applied from the opposite direction.
3 Remove all the old grease from the hub and bearings, giving the latter a final wash in petrol. Check the bearings for play or any signs of roughness as they are rotated. If there is any doubt about their condition, renew them.
4 Before driving the bearings back into the hub, pack the hub with new grease and also grease the bearings. Use the same double-diameter drift to drive them back into position, not forgetting the distance collar between them. Refit any oil seals or dust covers which have been displaced, renewing any that are damaged. Replace and tighten the left hand bearing retainer, then replace the speedometer drive gearbox assembly and the front brake disc, using new tab washers.

8 Front wheel - reassembly and replacement

1 Replace the front wheel in the forks and tighten the end caps so that the wheel spindle is held rigidly. A raised portion around the spindle engages with a channel in the end cap, to ensure correct location. Some models have a variation of the wheel spindle arrangement which has a locking nut on the left hand side of the forks.
2 Reconnect the speedometer cable and make sure that the cable is routed correctly through the wire clips on the mudguard to ensure that the cable does not come into contact with the tyre.
3 Spin the wheel to make sure that it revolves freely, then check the brake action. Make a final check that all nuts and bolts have been tightened fully.

9 Rear wheel - examination, removal and renovation

1 Place the machine on the centre stand so that the rear wheel is raised clear of the ground. Check for rim alignment, damage to the rim and loose or broken spokes by following the procedure relating to the front wheel, as described in Section 2 of this Chapter.
2 To remove the rear wheel, use the procedure described in paragraphs 2 to 4 of Section 9, Chapter 4. The rear brake plate and brake assembly can then be lifted off the right hand end of the hub as a complete unit.
3 The rear wheel bearings have a similar arrangement to those of the front wheel and can be removed and replaced in a similar manner. In this instance there is less difficulty in gaining access because there is no speedometer drive gearbox or brake disc to remove first. It will, however, be necessary to detach the rear wheel sprocket before access is available to the inner left hand bearing retainer. It is retained by four bolts which pass through the mounting flange. Bend back the tab washers and unscrew the four nuts, then lift the sprocket off the bolts. When the sprocket has been detached, the flange on which the sprocket was mounted can be lifted away from the cush drive assembly within the hub. The bearing retainer is now exposed and can be unscrewed, to permit removal of the bearings.
4 Note that there is an outer left hand bearing within the centre of the sprocket mounting flange. If the retainer in front of this bearing is unscrewed, this third bearing complete with oil seal and spindle sleeve can be drifted out from the cush drive end of the flange.

10.1. Brake plate with brake assembly will lift off from rear wheel hub

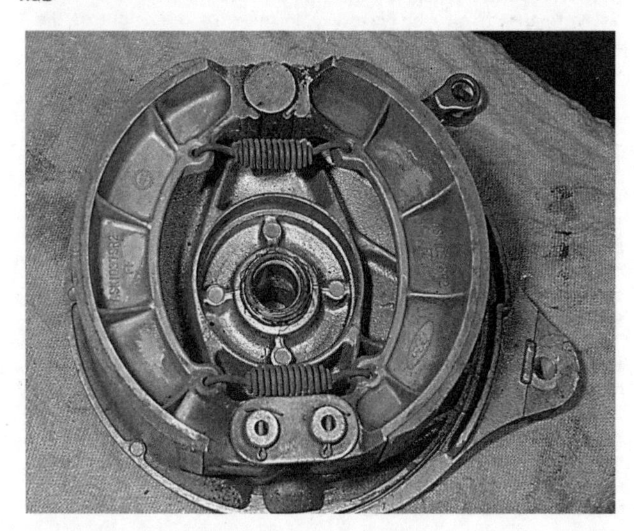

10.3. Split pins and anchor plate must be removed to release brake shoes

10 Rear brake assembly - examination, renovation and reassembly

1 The rear brake is a single leading shoe of conventional design. The brake assembly is exposed when the brake plate is lifted off the wheel hub from the right hand side.
2 Examine the condition of the brake linings. If they are thin or uneven, the brake shoes should be renewed. The linings are bonded on and cannot be supplied separately.
3 To remove the brake shoes, turn the brake operating lever so that the brake is in the fully on position. Pull the brake shoes apart against their return spring pressure to free them from the operating cam after withdrawing the split pins and plates used to anchor their ends. The shoes can then be lifted away, complete with the return springs by reverting to a V formation. When they are clear of the brake plate, the return springs can be removed and the shoes separated.
4 Before replacing the brake shoes, check that the brake operating cam is working smoothly and not binding in the pivot bush. The cam is removed for greasing by detaching the operating arm from the splined shaft, after first marking its position in relation to the splines so that it is replaced in an identical position. Slacken the pinch bolt and draw the arm off the shaft. Grease the cam and spindle sparingly, or grease may work onto the brake linings and impair their efficiency.
5 Check the inner surface of the brake drum. The surface on

which the brake shoes operate should be smooth and free from score marks or indentations, otherwise reduced braking efficiency will be inevitable. Remove all traces of brake lining dust and wipe with a clean rag soaked in petrol to remove all traces of grease and oil.
6 To reassemble the brake shoes on the brake plate, fit the return springs and pull the shoes apart, holding them in V formation. If they are now located with the brake operating cam and pivots, they can be pushed back into position by pressing downward in order to snap them into position. Do not use excessive force, or there is risk of distorting the brake shoes permanently. Do not forget to replace the anchor plates and split pins which retain the shoes in position.

11 Adjusting the rear brake

1 If the adjustment of the rear brake is correct, the brake pedal will have a travel of from 20 to 30 mm (0.8 to 1.2 inch). Adjustment is made at the end of the brake operating rod, at the point where the rod passes through the trunnion in the brake operating arm. To decrease pedal movement, screw the adjusting nut inward and outward for increase of travel.
2 Note that it may be necessary to re-adjust the height of the stop lamp switch if the range of pedal travel has been altered to any marked extent.

12 Cush drive assembly - examination and renovation

1 The cush drive assembly is contained in the left hand side of the rear wheel hub. It takes the form of four rectangular blocks cast into the rear of the flange on which the rear sprocket is mounted. These blocks abut against a series of rubbers (eight in total) contained with the hub of the rear wheel and permit the sprocket to move in relation to the wheel within certain limits. This action effectively cushions any surges or roughness in the transmission, which would otherwise convey the impression of harshness.
2 When the rear wheel is removed, it is advisable to examine the rubbers for signs of damage or general deterioration which may otherwise render them ineffective. The usual sign of a cush drive assembly requiring attention takes the form of excessive sprocket movement.

13 Rear wheel sprocket - removal, examination and replacement

1 Removal of the rear sprocket is accomplished by following the procedure described in Section 9 of this Chapter, paragraphs 1 to 3.
2 The sprocket will have to be renewed if the teeth are badly worn or if they are hooked, chipped or broken. It is bad practice, however, to renew one sprocket on its own. The final drive sprockets should always be renewed as a pair and a new chain fitted, otherwise more rapid wear will take place, necessitating even earlier replacement.
3 No advantage is gained from varying the size of either sprocket. The sizes selected have been chosen by the manufacturer as the result of exhaustive tests, to give optimum performance with the existing engine characteristics.
4 Sprocket sizes have been varied by the manufacturer since the inception of the original CB 750 model, but always in pairs to avoid any sharp change in ratios. Early models have a 16 tooth gearbox sprocket and a 45 tooth rear wheel sprocket, an arrangement which encouraged a high tensional loading on the chain. The problem was overcome by reverting to a 12 tooth gearbox sprocket and a 48 tooth rear wheel sprocket, which has greatly improved chain life. These are the sizes which should be fitted when the original sprockets are renewed on early models. Always use new tab washers when fitting sprocket retaining nuts.

10.6. Do not omit split pin through torque arm connection, when replacing wheel

14.1. Centre of gearbox sprocket contains adjustable chain oiler

14.4. Chainstays have graduation marks to aid accurate wheel alignment

14 Final drive chain - examination and lubrication

1 The final drive chain is fully exposed apart from the protection given by a short chainguard along the upper run. Some lubrication is given by a controlled leak of oil along the final drive shaft which distributes oil to the chain by centrifugal force across the face of the gearbox sprocket. This, however, is no substitute for regular maintenance, even though the rate of feed can be controlled by the small adjusting screw in the centre of the sprocket.

2 Irrespective of the amount of protection provided, the chain tension will need adjustment at regular intervals, to compensate for wear. This is accomplished by slackening the rear wheel nut, after withdrawing the split pin, when the machine is on the centre stand. The draw bolt adjusters in each rear fork end are used to draw the wheel backward until the chain is again under correct tension. Note that it may be necessary to slacken the rear brake torque arm bolt during this operation.

3 Chain tension is correct if there is from 15 to 20 mm (0.6 to 0.8 inch) of slack in the middle of the lower run. Always check the chain at its tightest point; a chain rarely wears in an even manner during service.

4 Always adjust the draw bolts an even amount so that correct wheel alignment is preserved. The fork ends are marked with a series of vertical lines to provide a visual check. If desired, wheel alignment can be checked by rubbing a plank of wood parallel to the machine so that it touches both walls of the rear tyre. If wheel alignment is correct, it should be equidistant from either side of the front wheel tyre when tested on both sides of the rear wheel; it will not touch the front tyre because this tyre has a smaller cross section. See the accompanying diagram.

5 Do not run the chain overtight to compensate for uneven wear. A tight chain will place excessive stresses on the gearbox and rear wheel bearings leading to their early failure. It will also absorb a surprising amount of power.

6 After a period of running, the chain will require lubrication. Lack of oil will accelerate the rate of wear of both chain and sprockets and will lead to harsh transmission. The application of engine oil will act as a temporary expedient, but it is preferable to remove the chain and immerse it in a molten lubricant such as Linklyfe or Chainguard after it has been cleaned in a paraffin bath. These latter lubricants achieve better penetration of the chain links and rollers and are less likely to be thrown off when the chain is in motion.

7 To check whether the chain is due for replacement, lay it lengthwise in a straight line and compress it endwise until all play is taken up. Anchor one end, then pull in the opposite direction to take up the play which develops. If the chain extends by more than ¼ inch per foot, it should be renewed in conjunction with the sprockets. Note that this check should ALWAYS be made after the chain has been washed out, but before any lubricant is applied, otherwise the lubricant may take up some of the play.

8 When fitting the chain on the machine, make sure the spring link is positioned correctly with the closed end facing the direction of travel.

9 Late models have an endless chain and it is not advisable to substitute a spring link if the chain has to be separated.

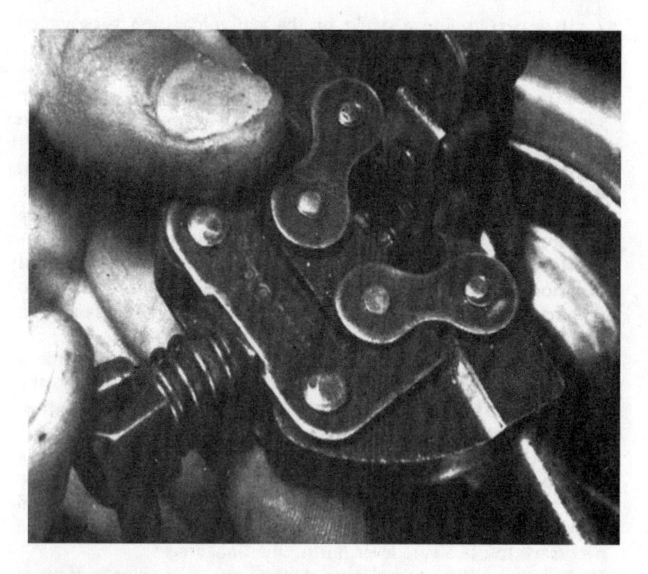

14.9. Always use a chain rivet extractor to separate an endless chain

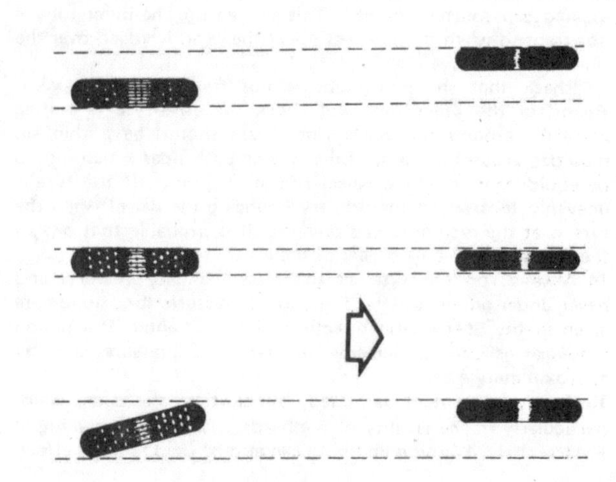

Fig. 5.4. Method of checking wheel alignment

15 Tyres - removal and replacement

1 At some time or other the need will arise to remove and replace the tyres, either as the result of a puncture or because a renewal is required to offset wear. To the inexperienced, tyre changing represents a formidable task yet if a few simple rules are observed and the technique learned, the whole operation is surprisingly simple.

2 To remove the tyre from either wheel, first detach the wheel from the machine by following the procedure in Chapters 4.2, paragraph 7 or 4.9, paragraphs 2 to 4, depending on whether the front or the rear wheel is involved. Deflate the tyre by removing the valve insert and when it is fully deflated, push the bead of the tyre away from the wheel rim on both sides so that the bead enters the centre well of the rim. Remove the locking cap and push the tyre valve into the tyre itself.

3 Insert a tyre lever close to the valve and lever the edge of the tyre over the outside of the wheel rim. Very little force should be necessary; if resistance is encountered it is probably due to the fact that the tyre beads have not entered the well of the wheel rim all the way round the tyre.

4 Once the tyre has been edged over the wheel rim, it is easy to work around the wheel rim so that the tyre is completely free on one side. At this stage, the inner tube can be removed.

5 Working from the other side of the wheel, ease the other edge of the tyre over the outside of the wheel rim which is furthest away. Continue to work around the rim until the tyre is free completely from the rim.

6 If a puncture has necessitated the removal of the tyre, re-inflate the inner tube and immerse it in a bowl of water to trace the source of the leak. Mark its position and deflate the tube. Dry the tube and clean the area around the puncture with a petrol soaked rag. When the surface has dried, apply the rubber solution and allow this to dry before removing the backing from the patch and applying the patch to the surface.

7 It is best to use a patch of the self-vulcanising type which will form a very permanent repair. Note that it may be necessary to remove a protective covering from the top surface of the patch, after it has sealed in position. Inner tubes made from synthetic rubber may require a special type of patch and adhesive if a satisfactory bond is to be achieved.
8 Before refitting the tyre, check the inside to make sure that the agent which caused the puncture is not trapped. Check the outside of the tyre, particularly the tread area, to make sure nothing is trapped that may cause a further puncture.
9 If the inner tube has been patched on a number of past occasions, or if there is a tear or large hole, it is preferable to discard it and fit a new one. Sudden deflation may cause an accident, particularly if it occurs with the front wheel.
10 To replace the tyre, inflate the inner tube sufficiently for it to assume a circular shape but only just. Then push it into the tyre so that it is enclosed completely. Lay the tyre on the wheel at an angle and insert the valve captive in its correct location.
11 Starting at the point furthest from the valve, push the tyre bead over the edge of the wheel rim until it is located in the central well. Continue to work around the tyre in this fashion until the whole of one side of the tyre is on the rim. It may be necessary to use a tyre lever during the final stages.
12 Make sure that there is no pull on the tyre valve and again commencing with the area furthest from the valve, ease the other bead of the tyre over the edge of the rim. Finish with the area close to the valve, pushing the valve up into the tyre until the locking cap touches the rim. This will ensure the inner tube is not trapped when the last section of the bead is edged over the rim with a tyre lever.
13 Check that the inner tube is not trapped at any point. Re-inflate the inner tube and check that the tyre is seating correctly around the wheel rim. There should be a thin rib moulded around the wall of the tyre on both sides which should be equidistant from the wheel rim at all points. If the tyre is unevenly located on the rim, try bouncing the wheel when the tyre is at the recommended pressure. It is probable that one of the beads has not pulled clear of the centre well.
14 Always run the tyres at the recommended pressures and never under or over-inflate. The correct pressures for solo use are given in the Specifications section of this Chapter. If a pillion passenger is carried, increase the rear tyre pressure only by approximately 4 psi.
15 Tyre replacement is aided by dusting the side walls, particularly in the vicinity of the beads, with a liberal coating of French chalk. Washing up liquid can also be used to good effect, but this has the disadvantage of causing the inner surfaces of the wheel rim to rust.
16 Never replace the inner tube and tyre without the rim tape in position. If this precaution is overlooked there is good chance of the ends of the spoke nipples chafing the inner tube and causing a crop of punctures.
17 Never fit a tyre which has a damaged tread or side walls. Apart from the legal aspects, there is a very great risk of a blow-out, which can have serious consequences on any two-wheel vehicle.
18 Tyre valves rarely give trouble, but it is always advisable to check whether the valve itself is leaking before removing the tyre. Do not forget to fit the dust cap, which forms an effective second seal. This is especially important on high performance machines where centrifugal force can cause the valve insert to retract and the tyre to deflate without warning.

16 Front wheel - balancing

1 It is customary on all high performance machines to balance the front wheel complete with tyre and tube. The out of balance forces which exist are eliminated and the handling of the machine is improved in consequence. A wheel which is badly out of balance produces through the steering a most unpleasant hammering effect at high speeds.
2 Some tyres, including those fitted to the Honda CB 750 models as standard, have a balance mark on the sidewall, usually in the form of a coloured dot. This mark must be in line with the tyre valve, when the tyre is fitted to the inner tube. Even then, the wheel may require the addition of balance weights, to offset the weight of the tyre valve itself.
3 If the front wheel is raised clear of the ground and is spun, it will probably come to rest with the tyre valve or the heaviest part downward and will always come to rest in the same position. Balance weights must be added to a point diametrically opposite this 'heavy' spot until the wheel will come to rest in ANY position after it is spun.
4 Balance weights which clip around the wheel spokes are normally available in 5, 10 or 20 gramme sizes. If they are not available, wire solder, wrapped around the spokes close to the spoke nipples, form a good substitute.
5 There is no necessity to balance the rear wheel under normal road conditions, although it is advisable to replace the rear wheel tyre so that any balance mark is in line with the tyre valve.

Tyre changing sequence - tubed tyres

Deflate tyre. After pushing tyre beads away from rim flanges push tyre bead into well of rim at point opposite valve. Insert tyre lever adjacent to valve and work bead over edge of rim.

Use two levers to work bead over edge of rim. Note use of rim protectors

Remove inner tube from tyre

When first bead is clear, remove tyre as shown

When fitting, partially inflate inner tube and insert in tyre

Work first bead over rim and feed valve through hole in rim. Partially screw on retaining nut to hold valve in place.

Check that inner tube is positioned correctly and work second bead over rim using tyre levers. Start at a point opposite valve.

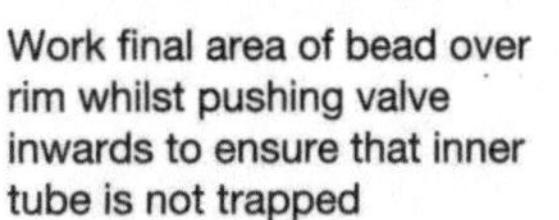

Work final area of bead over rim whilst pushing valve inwards to ensure that inner tube is not trapped

17 Fault diagnosis

Symptom	Reason/s	Remedy
Handlebars oscillate at low speeds	Buckle or flat in wheel rim, most probably front wheel	Check rim alignment by spinning wheel. Correct by retensioning spokes or rebuilding on new rim.
	Tyre not straight on rim	Check tyre alignment.
Machine lacks power and accelerates poorly	Rear brake binding	Warm brake drum provides best evidence. Re-adjust brake.
Rear brake grabs when applied gently	Ends of brake shoes not chamfered	Chamfer with file.
	Elliptical brake drum	Lightly skim in lathe (specialist attention required).
Front brake feels spongy	Air in hydraulic system	Bleed brake.
Brake pull-off sluggish	Brake cam binding in housing	Free and grease.
	Weak brake shoe springs	Renew if springs have not become displaced.
	Sticking pistons in brake caliper	Overhaul caliper unit.
Harsh transmission	Worn or badly adjusted final drive chain	Adjust or renew as necessary.
	Hooked or badly worn sprockets	Renew as a pair.
	Worn or deteriorating cush drive rubbers	Renew rubbers.

Chapter 6 Electrical system

Contents

Specifications

Battery

Make	Yuasa
Type	B64-12
Voltage	12 volts
Amp/hour capacity	14
Earth	Negative

Generator

Make	Hitachi
Output	13 amps

Ignition coils

Make	Toyo Denso

Starter

Brush length	0.427 - 0.512 in (12 - 13 mm)
Minimum length	0.217 in (5.5 mm)

Regulator unit

Make	Hitachi
Type	TLIZ-38
Core gap	0.024 - 0.04 in (0.6 - 1.0 mm)
Points gap	0.012 - 0.016 in (0.3 - 0.4 mm)

Silicon rectifier

Make	Hitachi
Type	SB6B-7
Output	12 volts, 15 amps

Bulbs

Headlamp	50/40W Sealed beam (late models)
Tail/stop	7/23W Offset pins
Pilot lamp	6W
Instrument lamps	3W
Indicator lamps	3W
Flashing indicator lamps	25W

All bulbs rated at 12 volts

Flashing indicator relay	Signal-Stat 142

1 General description

The Honda CB 750 is fitted with a 12 volt electrical system. The system comprises a crankshaft driven AC generator of the excited field, 3 phase type, the output of which is controlled by an electro-mechanical regulator to match the electrical demand. A silicon rectifier is incorporated in the circuit to convert the current to DC so that it can be used to charge the battery. The daytime charge rate does not normally exceed 2 - 3 amps at low rpm, but under night riding conditions when the lighting set is in full operation, this rate rises to 6.5 amps.

2 Crankshaft generator - checking the output

1 As explained in Chapter 3.2, the output from the generator can be checked by connecting both an ammeter and a voltmeter into the battery circuit.
2 Note that the test described gives only an approximate indication of whether the generator is functioning correctly, within its prescribed limits. It will be necessary to seek the assistance of either a Honda agent or an auto-electrician to determine whether the generator is working at peak efficiency. Note that a faulty regulator unit can give the appearance of a malfunctioning generator.
3 The generator field and stator coils can be checked without the need for any dismantling. Disconnect the block connector from the generator casing and identify the wire colours on the generator side of the connector. To check the field coil, measure the resistance between the white wire and the green wire. A reading of approximately 7.2 ohms should be obtained. The stator coils can be checked by measuring the resistance between each pair of yellow wires (3 tests). In this instance the result should be approximately 0.2 ohm.
4 If the results obtained differ considerably from the specifications, or if an open or short circuit is indicated, the coils will require renewal. It is, however, worth having your findings verified by a Honda agent prior to renewing the coils.

3 Battery - examination and maintenance

1 A Yuasa B64-12, 12 volt, 14 amp hour capacity battery is fitted to the Honda CB 750. It is mounted in the electrical compartment under the seat and has a negative earth connection.
2 The transparent plastics case of the battery permits the upper and lower levels of the electrolyte to be observed without disturbing the battery by removing the left hand side cover. Maintenance is normally limited to keeping the electrolyte level between the prescribed upper and lower limits and making sure that the vent tube is not blocked. The lead plates and their separators are also visible through the transparent case, a further guide to the general condition of the battery.
3 Unless acid is spilt, as may occur if the machine falls over, the electrolyte should always be topped up with distilled water to restore the correct level. If acid is spilt onto any part of the machine, it should be neutralised with an alkali such as washing soda or baking powder and washed away with plenty of water, otherwise serious corrosion will occur. Top up with sulphuric acid of the correct specific gravity (1.260 to 1.280) only when spillage has occurred. Check that the vent pipe is well clear of the frame or any of the other cycle parts.
4 It is seldom practicable to repair a cracked battery case because the acid present in the joint will prevent the formation of an effective seal. It is always best to renew a cracked battery, especially in view of the corrosion which will be caused if the acid continues to leak.
5 If the machine is not used for a period, it is advisable to remove the battery and give it a 'refresher' charge every six weeks or so from a battery charger. If the battery is permitted to discharge completely, the plates will sulphate and render the battery useless.
6 Occasionally, check the condition of the battery terminals to ensure that corrosion is not taking place and that the electrical connections are tight. If corrosion has occurred, it should be cleaned away by scraping with a knife and then using emery cloth to remove the final traces. Remake the electrical connections whilst the joint is still clean, then smear the assembly with petroleum jelly (NOT grease) to prevent recurrence of the corrosion. Badly corroded connections can have a high electrical resistance and may give the impression of a complete battery failure.

4 Battery - charging procedure

1 Since the ignition system is dependent on the battery for its operation, if the battery discharges completely it must be removed and recharged before the machine can be used. A battery charger is necessary for this purpose.
2 The normal charge rate is 4 amps for about 4 hours for a 14 amp hour battery. A more rapid charge at a higher rate can be given in an emergency, but this should be avoided if at all possible because it will shorten the useful working life of the battery. Always ensure the battery is topped up before charging.
3 When the battery is replaced on the machine, make sure that it is protected by the rubber pads in the battery compartment, which help damp out the undesirable effects of vibration. Do not reverse connect the battery, or the silicon rectifier may be damaged by the reverse flow of current.

5 Silicon rectifier - general description

1 The function of the silicon rectifier is to convert the AC current produced by the alternator into DC so that it can be used to charge the battery.
2 The rectifier is located to the rear of the battery, beneath the dual seat, a location where it is afforded reasonable protection. The question of access is of relatively little importance because the rectifier is unlikely to give trouble during normal service. Should it malfunction, a repair is not practicable. It must be renewed.
3 Damage to the rectifier will occur if the machine is run without a battery for any period of time, or with one that no longer holds its charge. A high voltage will develop in the absence of any load across the coils of the alternator which will cause a reverse flow of current and subsequent damage to the rectifier cells. Reverse connection of the battery will have a similar undesirable effect.
4 There is no simple means of checking whether the rectifier is functioning correctly without the appropriate test equipment. A Honda agent or an auto-electrician are best qualified to advise, particularly if the battery is in a low state of charge.
5 Do not disturb the rectifier retaining nut or in any way damage the surfaces of the assembly. Any such action may cause the coating over the electrodes to peel or flake and destroy the working action.

6 Voltage regulator - examination and adjustment

1 The regulator unit is located within the electrical compartment, in close proximity to the starter solenoid. Access is gained by removing the left hand side cover. The cover of the regulator unit is retained in position by two screws.
2 To check the regulator unit, remove the cover, then connect up a voltmeter and ammeter exactly as described in Chapter 3, Section 2.2. Do not forget to ensure the battery is fully charged before commencing the check. Start the engine and if the charge readings are low (or too high) when compared with the engine rpm v charge rate readings listed in Section 2.3, the regulator requires adjustment.
3 To increase the charge rate or battery voltage, slacken the voltage adjusting screw locknut and turn the adjusting screw clockwise. If the charge rate or battery voltage is too high, turn the screw in the opposite direction. Turn the screw only a small amount at a time before rechecking the voltmeter and ammeter readings. When the setting is correct, hold the screw steady and tighten the locknut. Check again that the reading is still correct.
4 Note that there will be a 0.5 volt rise in voltage as the low speed contacts change over to the high speed contacts in the regulator unit. If the change in voltage is greater than 0.5 volts or if there is a drop in voltage, the core gap probably requires adjustment. Stop the engine.
5 Before adjusting the core gap, check that the electrical contacts are not dirty or pitted. If they are, fine emery cloth can

be used to clean them without need for further dismantling. After the points have been checked, and if necessary, cleaned, use a feeler gauge to verify whether the core gap is within the recommended limits of 0.024 - 0.040 inch (0.6 - 1.0 mm). The gap is adjusted by means of the adjusting screw at the end opposite to that of the voltage adjusting screw. This screw has no locknut.

6 It is also advisable to check the condition of the other set of points and if necessary, clean them in the manner described in the preceding paragraph. The points gap must be within the range 0.012 - 0.016 inch (0.3 - 0.4 mm), measured by a feeler gauge. Adjust the gap by turning the screw located at right angles to the core gap screw.

7 Restart the engine and check again that all the readings are correct after all the above adjustments have been carried out. Stop the engine, refit all the original electrical connections, the cover of the regulator unit and the electrical compartment side cover.

8 If it is not possible to achieve the desired results, the regulator unit is at fault and should be renewed or checked by a Honda agent or an auto-electrician.

7 Fuses - location and replacement

1 The fuse holder is located behind the left hand side cover. It contains two spare 15 amp fuses and the 15 amp fuse necessary to protect the electrical equipment.

2 If a fuse blows, it should not be renewed until a check has shown whether a short circuit has occurred. This will involve checking the electrical circuit to identify and correct the fault. If this precaution is not observed, the replacement fuse, which may be the only spare, may blow immediately on connection.

3 When a fuse blows whilst the machine is running and no spare is available a 'get you home' remedy is to remove the blown fuse and wrap it in silver paper before replacing it in the fuse holder. The silver paper will restore electrical continuity by bridging the broken wire within the fuse. This expedient should never be used if there is evidence of a short circuit or other major electrical fault, otherwise more serious damage will be caused. Renew the 'doctored' fuse at the earliest possible opportunity to restore full circuit protection.

8 Starter motor - removal, examination and replacement

1 An electric starter motor, operated from a small push-button on the right hand side of the handlebars, provides an alternative and more convenient method of starting the engine, without having to use the kickstarter. The starter motor is mounted within a compartment at the rear of the cylinder block, closed by an oblong, chromium plated cover. Current is supplied from the battery via a heavy duty solenoid switch and a cable capable of carrying the very high current demanded by the starter motor on the initial start-up.

2 The starter motor drives a free running clutch immediately behind the generator rotor. The clutch ensures the starter motor drive is disconnected from the primary transmission immediately the engine starts. It operates on the centrifugal principle; spring loaded rollers take up the drive until the centrifugal force of the rotating engine overcomes their resistance and the drive is automatically disconnected.

3 To remove the starter motor from the engine unit, first disconnect the positive lead from the battery, then the starter motor cable from the solenoid switch. It is not possible to detach the starter cable from the starter motor itself; in consequence the cable must be threaded through the frame so that it is free when the starter motor is lifted away. Detach the two bolts in the chromium plated cover over the starter motor housing and lift the cover away, complete with gasket. The starter motor is secured to the crankcase by two bolts which pass through the left hand end of the motor casting. When these bolts are withdrawn, the motor can be prised out of position and lifted out of its compartment, with the heavy duty cable still

6.3. Turn voltage adjusting screw to vary charge rate

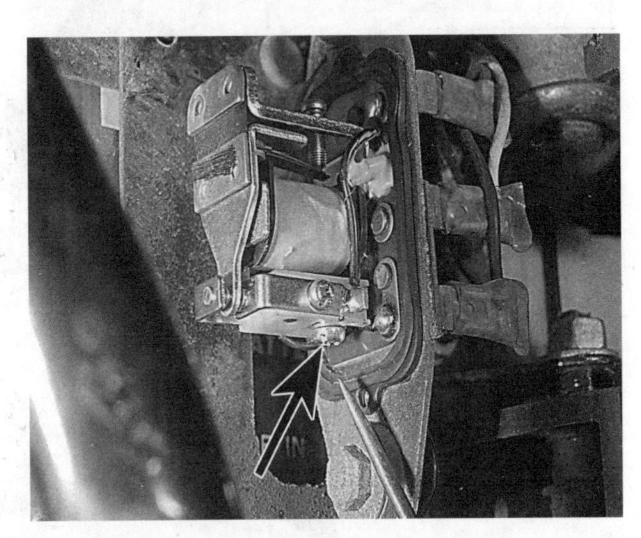

6.5. Adjust core gap if contact changeover is incorrect

6.6. Screw on side is for points gap adjustment

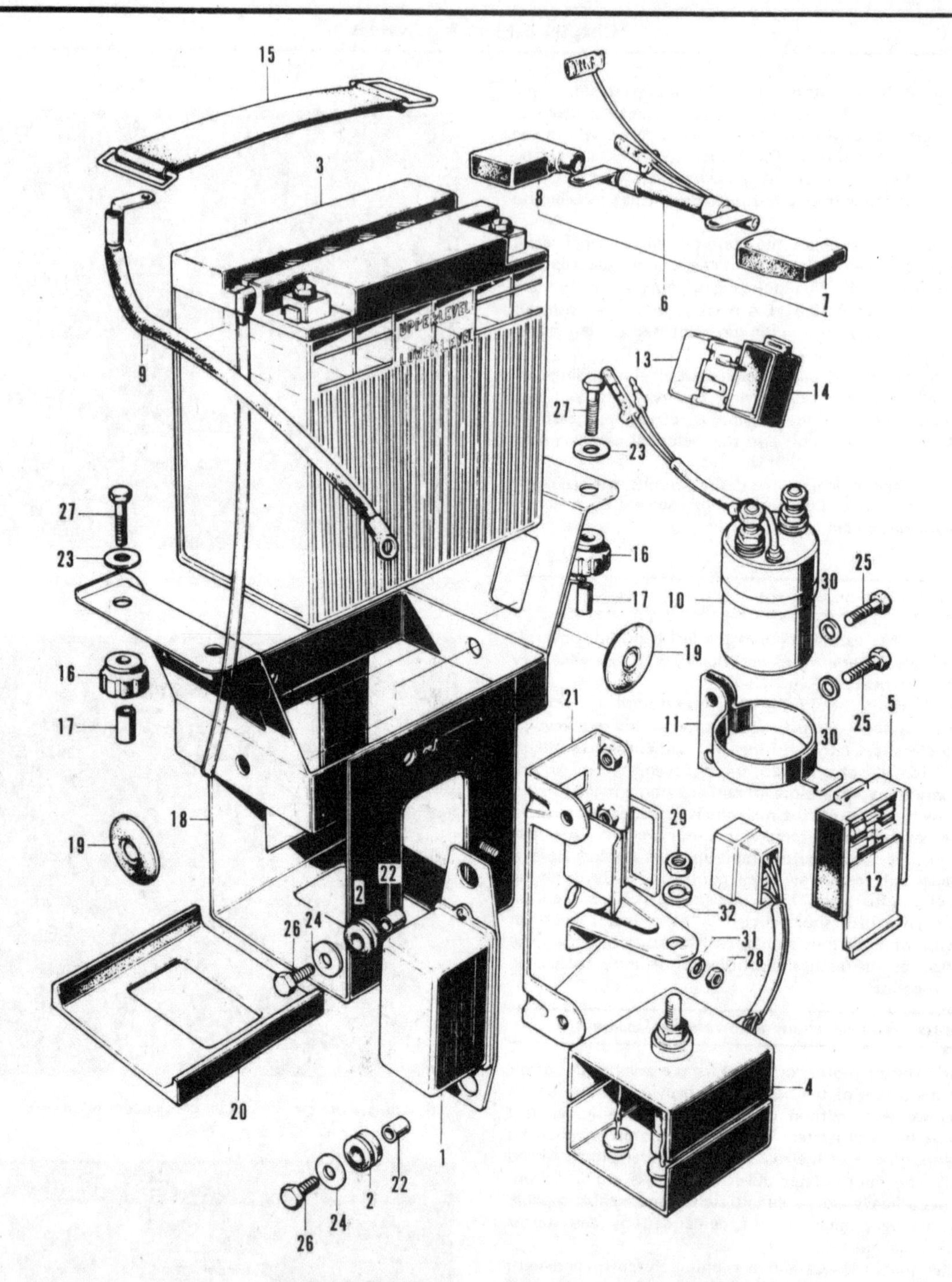

FIG. 6.1. BATTERY AND ELECTRICAL COMPONENTS

1	*Regulator (Hitachi)*	*17*	*Battery box collar - 4 off*
2	*Vibration insulating rubber - 2 off*	*18*	*Battery box*
3	*Battery (Yuasa B64-12)*	*19*	*Battery vibration insulating rubber - 4 off*
4	*Rectifier (Hitachi)*	*20*	*Battery vibration insulating rubber pad*
5	*Fuse box*	*21*	*Base plate for electrical components*
6	*Starter motor cable*	*22*	*Collar - 2 off*
7	*Battery terminal cover*	*23*	*Washer - 4 off*
8	*Battery terminal cover*	*24*	*Washer - 2 off*
9	*Battery earthing strap*	*25*	*Bolt - 2 off*
10	*Starter motor solenoid switch*	*26*	*Bolt - 2 off*
11	*Clamp for starter motor switch*	*27*	*Bolt - 4 off*
12	*Fuse (15 amp) - 4 off*	*28*	*Nut*
13	*Flashing indicator lamp relay*	*29*	*Nut*
14	*Rubber mounting for flashing indicator relay*	*30*	*Washer - 2 off*
15	*Battery strap*	*31*	*Spring washer - 6 mm*
16	*Battery mounting vibration insulator - 4 off*	*32*	*Spring washer - 8 mm*

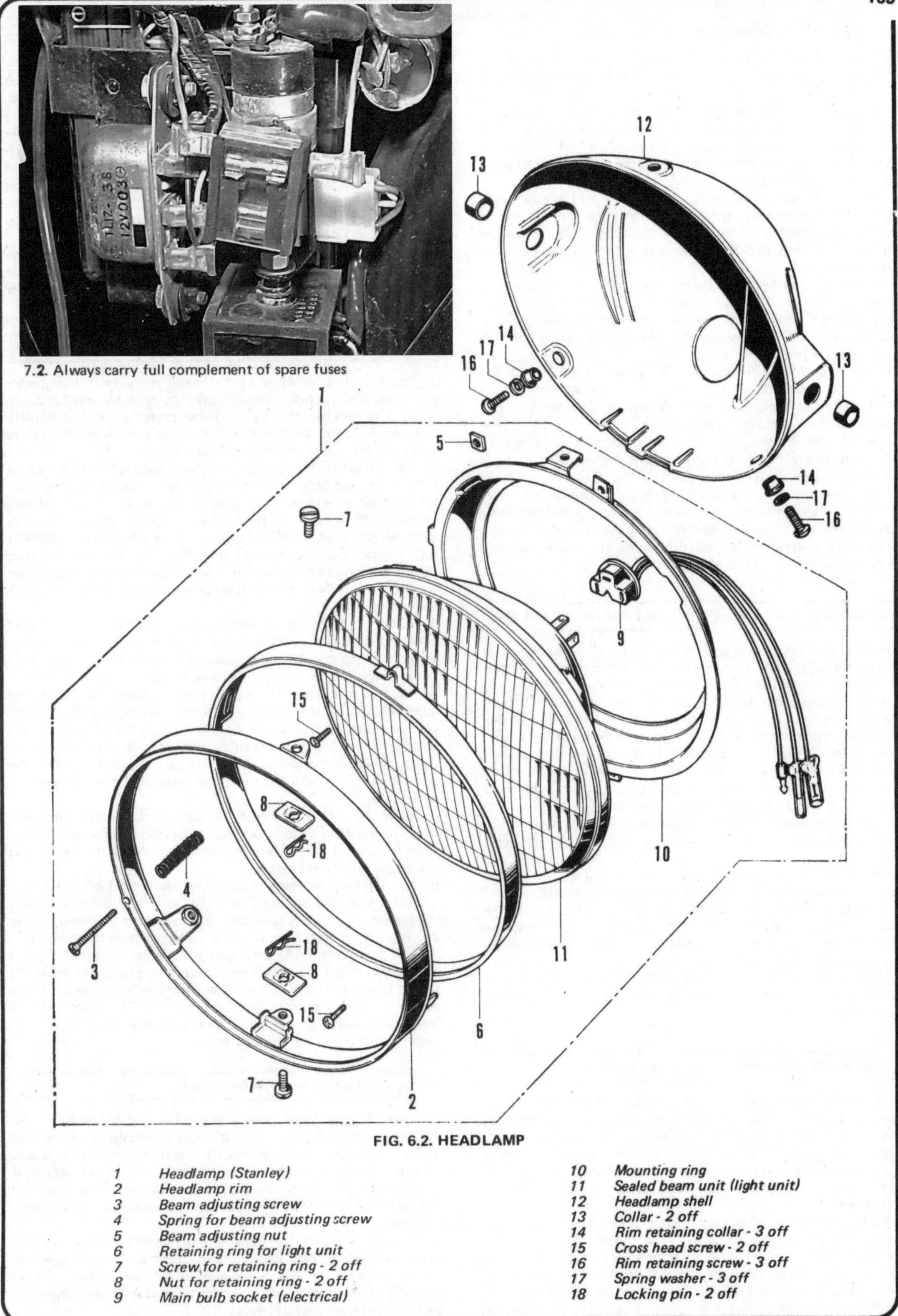

7.2. Always carry full complement of spare fuses

FIG. 6.2. HEADLAMP

1	*Headlamp (Stanley)*	10	*Mounting ring*
2	*Headlamp rim*	11	*Sealed beam unit (light unit)*
3	*Beam adjusting screw*	12	*Headlamp shell*
4	*Spring for beam adjusting screw*	13	*Collar - 2 off*
5	*Beam adjusting nut*	14	*Rim retaining collar - 3 off*
6	*Retaining ring for light unit*	15	*Cross head screw - 2 off*
7	*Screw for retaining ring - 2 off*	16	*Rim retaining screw - 3 off*
8	*Nut for retaining ring - 2 off*	17	*Spring washer - 3 off*
9	*Main bulb socket (electrical)*	18	*Locking pin - 2 off*

attached. If necessary, temporarily detach the lead and grommet from the oil pressure switch, which pass through the starter motor compartment.

4 The parts of the starter motor most likely to require attention are the brushes. The end cover is retained by the two long screws which pass through the lugs cast on both end pieces. If the screws are withdrawn, the end cover can be lifted away and the brush gear exposed.

5 Lift up the spring clips which bear on the end of each brush and remove the brushes from their holders. Each brush should have a length of 0.472 - 0.512 inch (12 - 13 mm). The serviceable limit is 0.217 inch (5.5 mm), which denotes when the brushes must be renewed.

6 Before the brushes are replaced, make sure the commutator is clean, on which they bear. Clean with a strip of fine emery cloth pressed against the commutator whilst the latter is revolved by hand. Wipe with a rag soaked in petrol to ensure a bright, grease-free surface is obtained. At the same time, check that the mica insulator between the copper segments of the armature is undercut. If the difference in height is less than 0.012 inch (0.3 mm) it must be re-cut - a task for a Honda dealer or an auto-electrical expert.

7 Replace the brushes in their holders and check that they slide quite freely. Make sure the brushes are replaced in their original positions because they will have worn to the profile of the commutator. Replace and tighten the end cover, then replace the starter motor and cable in the housing, tighten down and re-make the electrical connection to the solenoid switch. Check that the starter motor functions correctly before replacing the compartment cover and sealing gasket.

9 Starter motor free running clutch - construction and renovation

1 Although a mechanical and not an electrical component, it is appropriate to include the free running clutch in this Chapter because it is an essential part of the electric starter system.

2 As mentioned in Chapter 1, the free running clutch is built into the alternator rotor assembly and will be found in the back of the rotor when the latter is removed from the left hand end of the crankshaft. The only parts likely to require attention are the rollers and their springs, or the bush in the centre of the driven sprocket. Access to the rollers is gained by removing the three countersunk crosshead screws which retain the clutch body to the rear of the alternator rotor. Signs of wear or damage will be obvious and will necessitate renewal of the worn or damaged parts.

3 The bush in the centre of the driven sprocket behind the clutch will need renewal only after very extensive service.

4 To check whether the clutch is operating correctly, turn the driven sprocket anticlockwise. This should force the spring loaded rollers against the crankshaft and cause it to tighten on the crankshaft as the drive is taken up.

5 If the starter clutch has been dismantled, make sure the three crosshead screws are staked over after reassembly, to prevent them working loose.

10 Starter solenoid switch - function and location

1 The starter motor switch is designed to work on the electro-magnetic principle. When the starter motor button is depressed, current from the battery passes through windings in the switch solenoid and generate an electro-magnetic force which causes a set of contact points to close. Immediately the points close, the starter motor is energised and a very heavy current is drawn from the battery.

2 This arrangement is used for at least two reasons. Firstly, the starter motor current is drawn only when the button is depressed and is cut off again when pressure on the button is released. This ensures minimum drainage on the battery. Secondly, if the battery is in a low state of charge, there will not be sufficient current to cause the solenoid contacts to close. In consequence, it is not possible to place an excessive drain on the battery which, in some circumstances, can cause the plates to overheat and shed their coatings. If the starter will not operate, first suspect a discharged battery. This can be checked by trying the horn or switching on the lights. If this check shows the battery to be in good shape, suspect the starter switch which should come into action with a pronounced click. It is located under the dual seat, close to the battery, and can be identified by the heavy duty starter cable connected to it. It is not possible to effect a satisfactory repair if the switch malfunctions; it must be renewed.

3 On the K3 series of Honda CB 750 4's, check that the gearbox is in neutral before testing the components of the starter motor circuit. The circuit is arranged so that the starter cannot be operated until the gearbox is in the neutral position.

11 Headlamp - replacing bulbs and adjusting beam height

1 In order to gain access to the headlamp bulbs it is necessary first to remove the rim, complete with the reflector and headlamp glass. The rim is retained by three crosshead screws equally spaced around the headlamp shell. Remove the screws completely and draw the rim from the headlamp shell.

2 Early models have a main headlamp bulb which is a push fit into the central bulb holder of the reflector. The bulb holder can be replaced in one position only to ensure the bulb is always correctly focussed. It is retained by a spring under tension. A bulb of the twin filament type is fitted which has a 50/40W rating (machines imported to the UK). The pilot lamp bulb is bayonet fitting and fits within a bulb holder which has the same form of attachment to the headlamp reflector. This bulb has a 6W rating.

3 Later models have a sealed beam headlamp unit, rated at 50/40W with no provision for a pilot lamp except those imported into the UK and countries when this is a statutory requirement. If one filament blows, the complete unit must be renewed. To release the lamp unit, remove the horizontal adjusting screw, the upper and lower retaining lock pins and screws from the collar which clamps the light unit to the headlamp rim. Make a note of the setting of the adjusting screw, otherwise it will be necessary to re-adjust the beam height after installing the new light unit by reversing the dismantling procedure.

4 Beam height is adjusted by turning the adjusting screw fitted in the nine o'clock position when the headlamp is viewed from the front. Turn anticlockwise to lower the beam and clockwise to raise the beam.

5 UK lighting regulations stipulate that the lighting system must be arranged so that the light will not dazzle a person standing at a distance greater than 25 yards from the lamp, whose eye level is not less than 3 feet 6 inches above that plane. It is easy to approximate this setting by placing the machine 25 yards away from a wall, on a level road, and setting the beam height so that it is concentrated at the same height as the distance of the centre of the headlamp from the ground. The rider must be seated normally during this operation and also the pillion passenger, if one is carried regularly.

12 Stop and tail lamp - replacing bulbs

1 The tail lamp has a twin filament bulb of 7/23W rating, to illuminate the rear number plate and to indicate when the rear brake is applied. On some models the stop lamp also operates in conjunction with the front brake; a stop lamp switch is incorporated in the front brake cable to meet the statutory requirements of the country or state to which the machine is exported.

2 To gain access to the stop and tail lamp bulb, unscrew the two crosshead screws which retain the plastics lens cover in position. The bulb has a bayonet fitting and offset pins so that the stop lamp filament cannot be inadvertently connected with the tail lamp and vice versa.

11.2. Early models have a push-fit holder for main bulb

11.2a. Pilot lamp bulb has also a push-fit holder

12.1. Stop/tail lamp bulb is of twin filament type, with offset pins

12.1a. Front brake has separate stop lamp switch as part of hydraulic system

13 Flashing indicator lamps - replacing bulbs

1 Flashing indicator lamps are fitted to the front and rear of the machine. They are mounted on short 'stalks' through which the electrical leads pass. One set of indicators is attached to either side of the fork lug to which the headlamp is attached. The other set are attached to the rear mudguard, immediately beneath the end of the dual seat.
2 Each flasher unit has a bayonet fitting 25W bulb. To replace a bulb remove the plastics end cover which is retained by two crosshead screws. Models in the K3 series have twin filament bulbs which provide clearance lights when the headlamp is switched on.

14 Flashing indicator relay - location and replacement

1 The flashing indicator relay fitted in conjunction with the flashing indicator lamps is located with the other electrical components in the compartment under the dual seat. It is rubber mounted to isolate it from the harmful effects of vibration.
2 When the relay malfunctions it must be renewed; a repair is impracticable. When the unit is in working order, audible clicks will be heard which keep pace with the flash of the indicator lamps. If the lamps malfunction, check firstly that a bulb failure is not responsible, or the handlebar switch faulty. The usual symptom of a fault is one initial flash before the unit goes dead.
3 Take great care when handling a flashing indicator relay. It is easily damaged, if dropped.

15 Speedometer and tachometer heads - replacement of bulbs

1 Apart from the bulbs used to illuminate the dial of each instrument during the hours of darkness, the speedometer head and the tachometer head fitted to the earlier models contain the warning lamps which indicate when neutral is selected (green), when oil pressure is below the pre-set limit, when the flashing indicators are operating (amber) and when the headlamp is on main beam (red). All bulbs fitted to either of the instrument heads have identical bulb holders, which are a push fit into the base of the instrument case. Access is gained by slackening the chromium plated clamp around both instruments and lifting each upward so that the two screws can be removed from the underplate and the bulb holders exposed. It may be possible to remove the undercovers without need to raise the instruments, if the mounting angle permits good access to the crosshead screws.

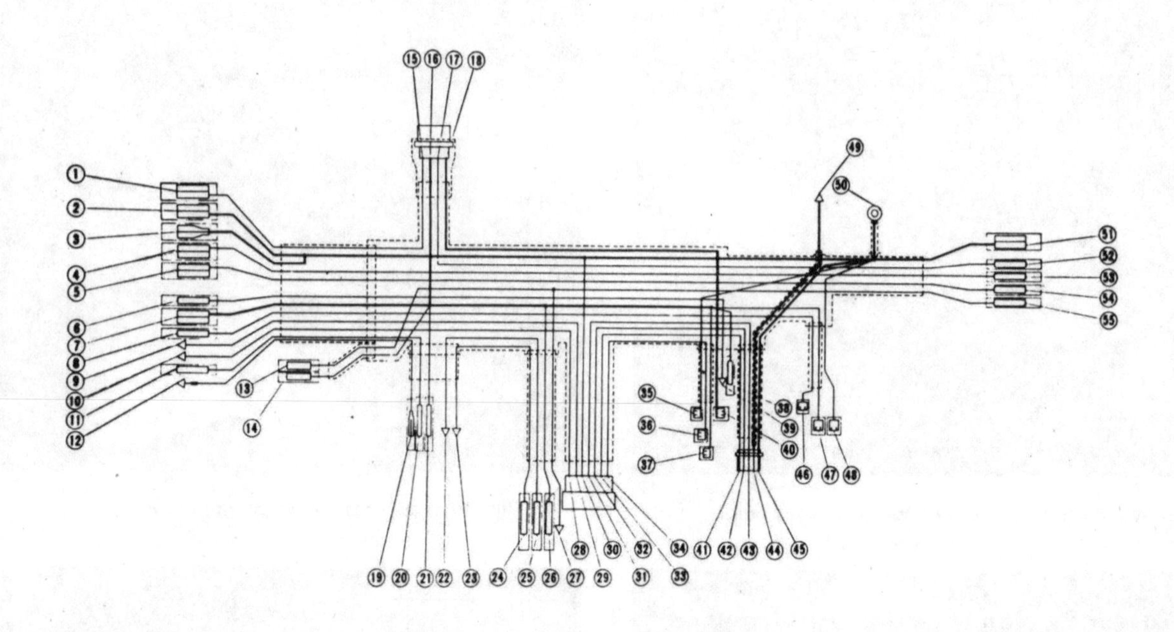

FIG. 6.3. COLOUR CODING OF WIRING HARNESS

No.	Lead colour	Connection
1	Brown/White	Speedometer lamp, Beam selector switch, Tachometer lamp
2	Green	Headlight, Speedometer, Tachometer
3	Green	R.L. front turn signal light
4	Light blue	R. front turn signal light, starter switch, turn signal pilot lamp
5	Orange	Turn signal pilot lamp, L. front turn single light, Starter/turn signal switch
6	Yellow/Red	Starter/turn signal switch
7	Black	Neutral pilot lamp, Oil pressure lamp, Head light beam selector switch, ignition switch
8	Grey (white tube)	Starter switch, Turn signal switch
9	Blue/red	Oil pressure lamp
10	Light green/Red	Neutral pilot lamp
11	Light green	Horn switch, Beam selector switch
12	Black/White	Ignition switch
13	Green/Yellow	Front stop switch
14	Black	Front stop switch
15	Brown/White	Main key switch
16	Black	" " "
17	Brown	" " "
18	Red	" " "
19	Black/White	Ignition coil
20	Light green	Horn
21	Black	Horn, ignition coil
22	Blue	Ignition coil
23	Yellow	Ignition coil
24	Yellow	Contact breaker
25	Blue	" "
26	Black	" "
27	Green/Yellow	Stop switch
28	Light green/Red	Neutral switch
29	Blue/Red	Oil pressure switch
30	Green	A.C. generator
31	Yellow	" "
32	Yellow	A.C. generator
33	"	" "
34	White	" "
35	Green	Regulator
36	White	"
37	Black	"
38	Black	Starter magnetic switch
39	Yellow/Red	" " "
40	Red	Fuse
41	Yellow	Silicone rectifier
42	"	" "
43	"	" "
44	Red/White	" "
45	Green	" "
46	Grey	Winker relay
47	Black	" "
48	Green	—
49	Red/White	Starter magnetic switch
50	Green	Frame body
51	Green	Tail/stop light, turn signal light
52	Brown	Tail light
53	Light blue	R. rear turn signal light
54	Orange	L. rear turn signal light
55	Green/Yellow	Stop light

16 Indicator panel lamps - replacement of bulbs

Late versions of the Honda CB 750 4's have an indicator panel clamped to the handlebar centre which contains the four warning lamps which were previously included in the tachometer and speedometer heads. The bulb holders are similar to those fitted in the tachometer and speedometer heads and can be detached from the base of the indicator panel by pulling them away.

17 Horn - adjustment

1 The horn is provided with an adjusting screw in the back of the horn body so that the sound volume can be varied, if necessary. To adjust the horn note, turn the screw not more than one half turn in either direction and test. If the note is weaker, or lost altogether, turn in the opposite direction. Continue adjusting by one half turn at a time until the desired volume and note is obtained.
2 The horn button is located on the left hand side of the handlebars. If the horn does not operate and an electrical meter shows no current is passing through the horn when the horn button is depressed, check the continuity of the light green coloured wire within the headlamp shell.

18 Ignition and lighting switch - removal and replacement

1 The main switch which controls both the ignition system and the lighting is attached to a clamp around the left hand of the two parallel upper frame tubes.
2 If the switch proves defective, it can be removed by unscrewing the two bolts which secure the bracket to the frame and separating the terminal connector at the end of the short wiring harness. To release the switch unit itself, unscrew the threaded ring around the outside of the lock.
3 Fit the new switch through the bracket, tighten the threaded ring, then bolt the switch bracket to the frame gusset. Reconnect the wires at the terminal connector. Remember that when a new switch is fitted, it will be necessary also to change the ignition key.

19 Stop lamp switches - adjustment

1 All models have a stop lamp switch fitted to operate in conjunction with the rear brake pedal. The switch is located immediately to the rear of the crankcase, on the right hand side of the machine. It has a threaded body, permitting a range of adjustment.
2 If the stop lamp is late in operating, slacken the locknuts and turn the body of the lamp in an anticlockwise direction so that the switch rises from the bracket to which it is attached. When the adjustment seems near correct, tighten the locknuts and test.
3 If the lamp operates too early, the locknuts should be slackened and the switch body turned clockwise so that it is lowered in relation to the mounting bracket.
4 As a guide, the light should operate after the brake pedal has been depressed by about 2 cm (¾ inch).
5 A stop lamp switch is also incorporated in the front brake cable, to give warning when the front brake is applied. This is not yet a statutory requirement in the UK, although it applies in many other countries and states.
6 The front brake stop lamp switch is built into the hydraulic system and contains no provision for adjustment. If the switch malfunctions, it must be renewed.

20 Handlebar switches - general

1 The arrangement of the handlebar switches will vary according to year of manufacture and the 'K' series number. Generally

14.1. Remove plastic lens to gain access to indicator lamp bulbs

15.1. Slacken clamp to release speedometer and tachometer heads

15.1a. Bulbs are a push fit in the underside of either instrument

speaking, the switches give little trouble, but if necessary they can be dismantled by separating the halves which form a split clamp around the handlebars. Note that the machine cannot be started until the ignition cut-out on the right hand end of the handlebars is turned to the central 'ON' position.

2 Always disconnect the battery before removing any of the switches, to prevent the possibility of a short circuit. Most troubles are caused by dirty contacts, but in the event of the breakage of some internal part, it will be necessary to renew the complete switch.

21 Fault diagnosis

Fault	Reason/s	Remedy
Complete electrical failure	Blown fuse	Check wiring and electrical components for short circuit before fitting new 15 amp fuse.
	Isolated battery	Check battery connections, also whether connections show signs of corrosion.
Dim lights, horn and starter inoperative	Discharged battery	Remove battery and charge with battery charger. Check generator output and voltage regulator settings.
Constantly blowing bulbs	Vibration or poor earth connection	Check security of bulb holders. Check earth return connections.
Starter motor sluggish	Worn brushes	Remove starter motor and renew brushes.
Parking lights dim rapidly	Battery will not hold charge	Renew battery at earliest opportunity.
Flashing indicators do not operate	Blown bulb	Renew bulb.
	Damaged flasher unit	Renew flasher unit.

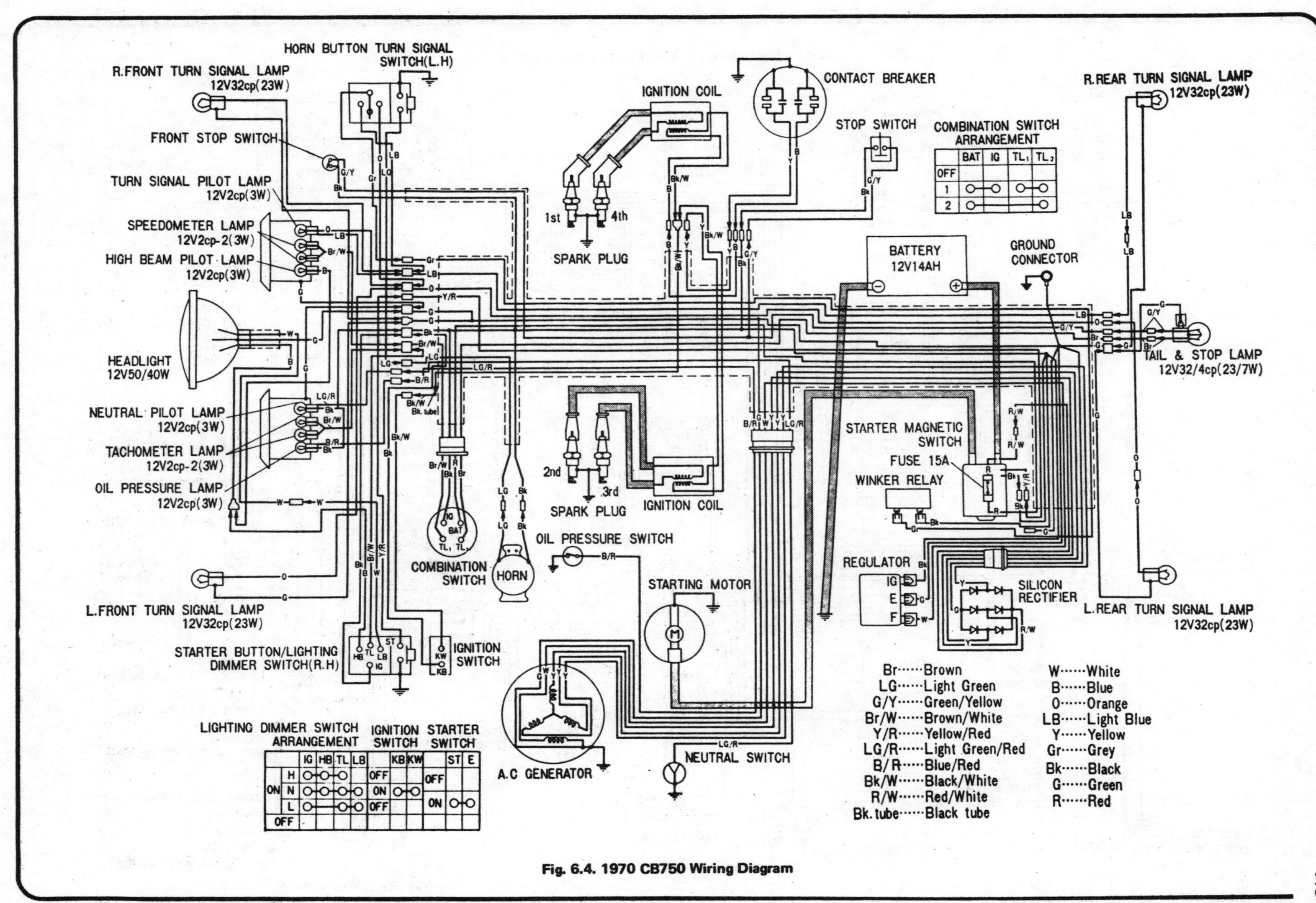

Fig. 6.4. 1970 CB750 Wiring Diagram

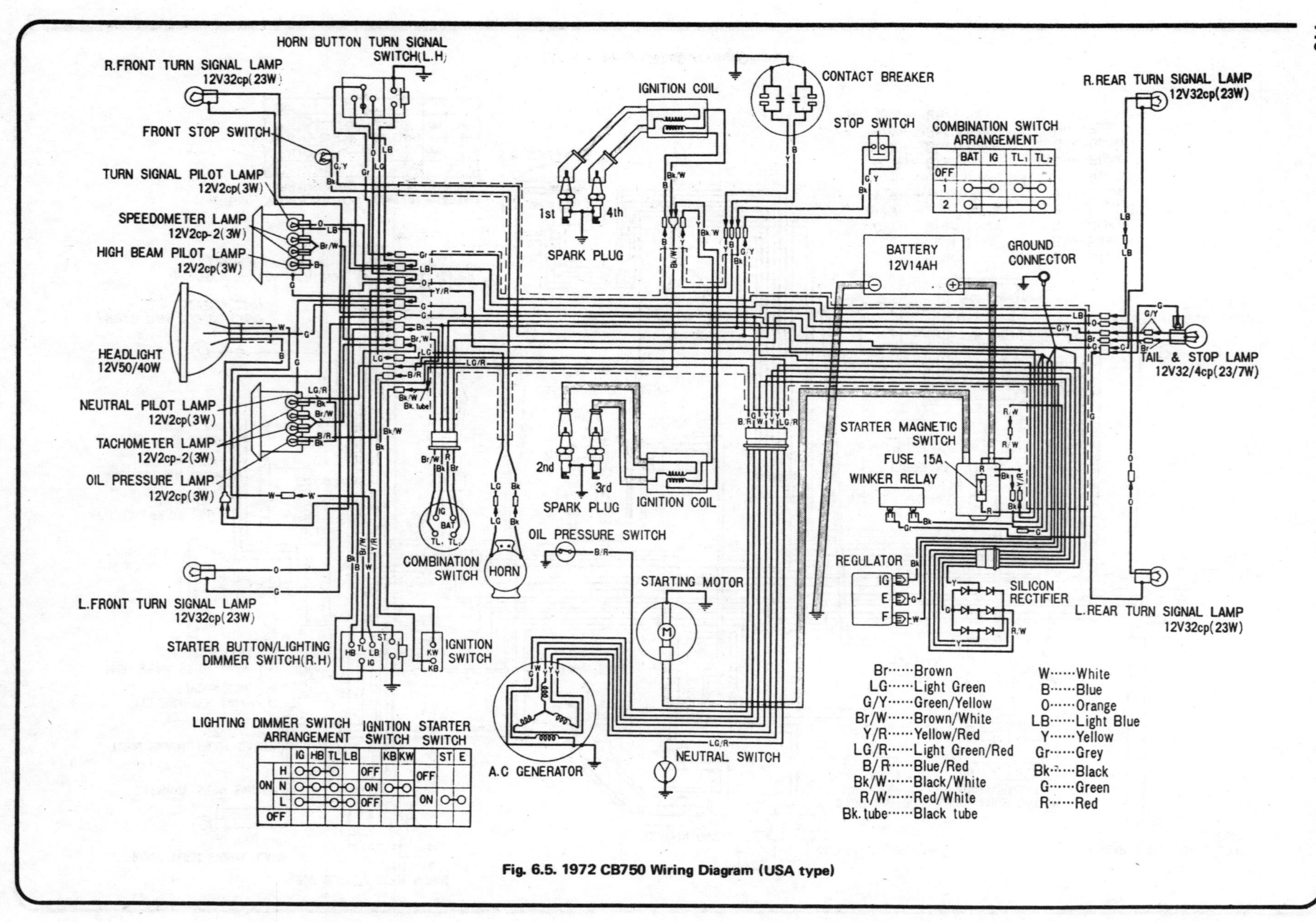

Fig. 6.5. 1972 CB750 Wiring Diagram (USA type)

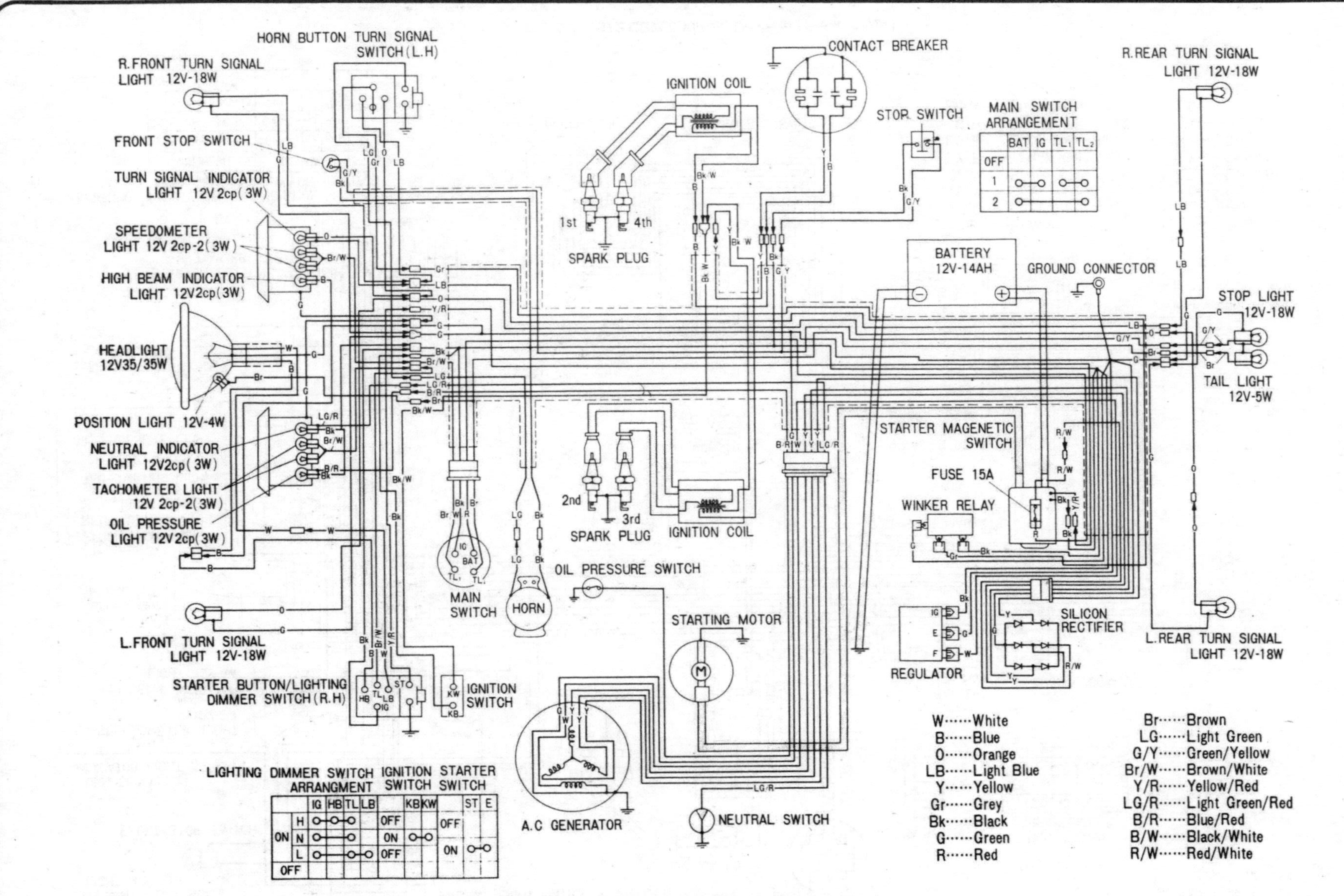

Fig. 6.6. 1972 CB750 Wiring Diagram (German type)

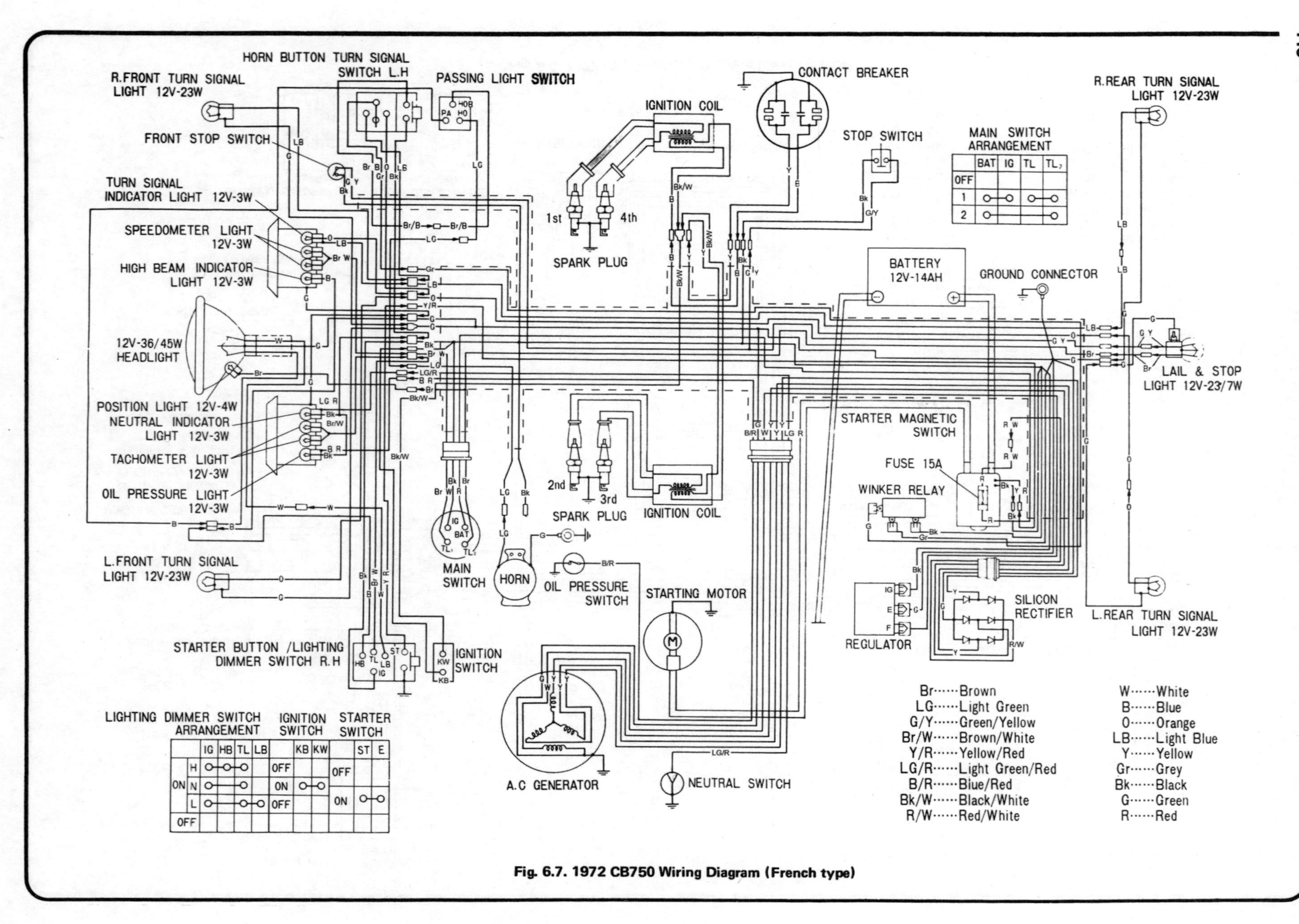

Fig. 6.7. 1972 CB750 Wiring Diagram (French type)

Chapter 7 The CB 750F model

Contents

1 General description

The CB750F model introduced in 1975, is in most respects similar to the 750 Fours which it supersedes. The obvious external differences are the 4 into 1 exhaust system and rear disc brake. Along with these go re-designed front forks and swinging arm. The front forks are now un-bushed, and have revised damping. The swinging arm pivot is simplified. The front brake hydraulic system is re-designed, but maintenance is unchanged. Minor changes are an engine breather system designed to reduce exhaust emission. The piston oil control ring is now of three part construction. Clutch design is slightly modified. The fuel tap is of different design.

The electrical system has significant changes, particularly on USA models. There are now three fuses, for main, headlight and tail light circuits respectively. The main switch has been moved to the top fork yoke, with a new warning lamp console. A safety feature incorporated is a clutch operated switch which, in conjunction with a diode and the neutral indicator switch, prevents operation of the starter motor if the machine is in gear with the clutch engaged. As indicated, USA models differ. They have a headlight which comes on when the main switch is turned on, and front indicators which remain on when main beam is selected. On USA models the steering lock is in the main switch, on UK models it is on the side of the steering head.

A novel feature, again on USA models, is a prop stand which retracts automatically if left extended, when the bike is ridden away.

2 Points to note

1 When replacing the cylinder head, note that there are now 8 dowels with rubber sealing sleeves on the inner studs, in addition to the two 'O' ring seals and two dowels previously fitted (Chapter 1, Section 36.1). Later cylinder head gaskets, incorporate a sealant which is squeezed out as the head is tightened down, making it very sticky. The recommended torque setting for the cylinder head nuts is 2.0 - 2.5 kg m (14.5 - 18.1 ft lb).

2 There are now oil restrictors with 'O' ring seals in the middle of the top of the cylinder head, supplying the rocker gear. Also two additional dowels at each end of the head (Chapter 1, Section 7).

3 The rear chain oiler is no longer fitted (Chapter 5, Section 14.1).

4 The recommended grade of spark plug is NGK D8ES-L or Nippon Denso X24ES-U.

5 The front brake caliper now has two hexagon headed clamp bolts, which are unscrewed from the outside of the caliper. Internally the caliper is the same. The caliper pivot is slightly modified, the disc spat is fitted to the pivot fixing bolts (Chapter 5, Section 4).

3 Exhaust system - removal

1 The four into one exhaust is retained by two bolts, fixing the silencer to the pillion footrest hanger, and should be removed as

3.2. Two bolts secure the exhaust

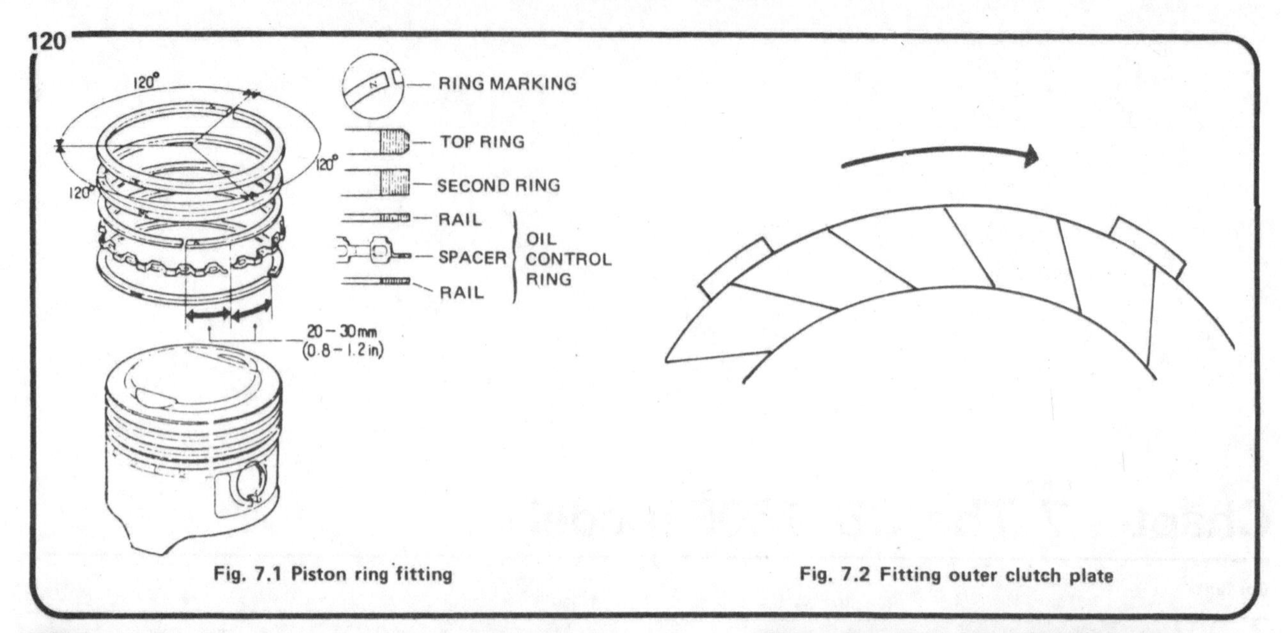

Fig. 7.1 Piston ring fitting

Fig. 7.2 Fitting outer clutch plate

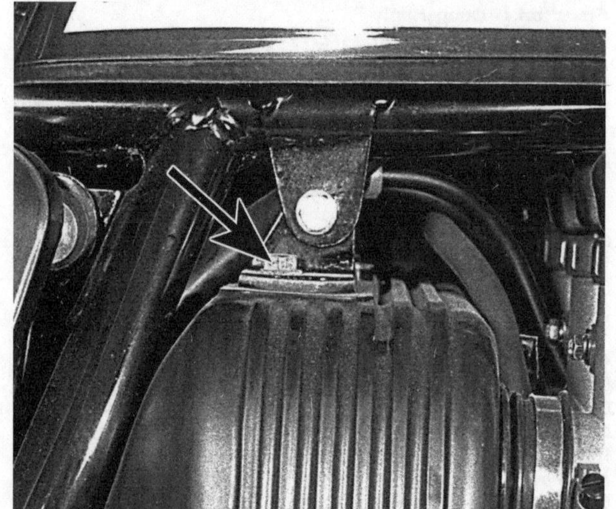

7.1 One bolt on each side fixes air filter bottom half

7.3 This bolt secures the engine breather filter

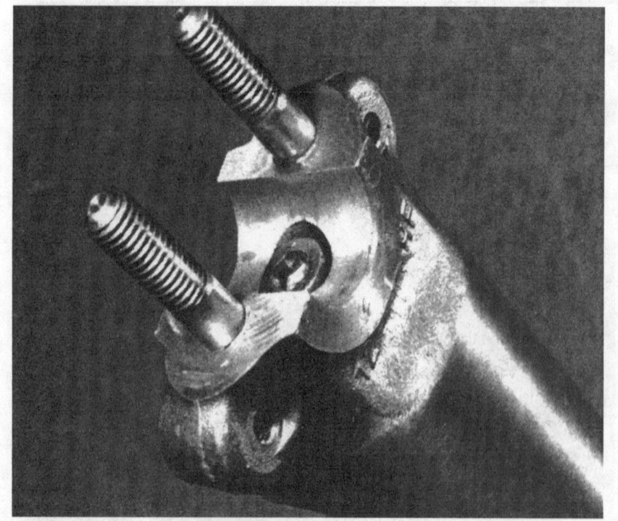

9.3 Unscrew socket head screw in end of fork stanchion

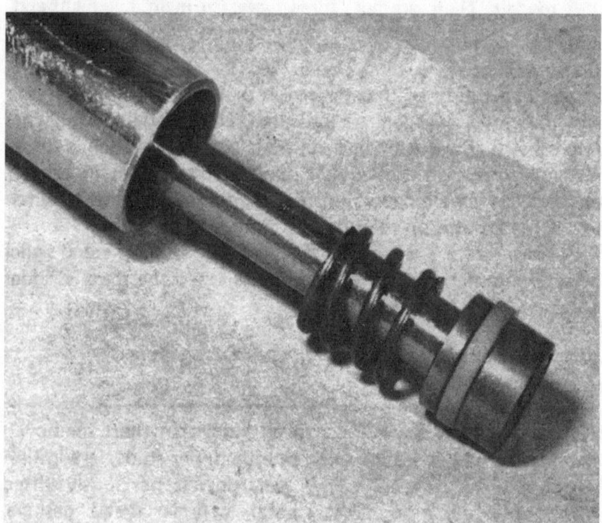

9.5 Push the damper out through top of fork stanchion

one assembly. The exhaust pipe to cylinder head fixing is also revised.
2 Unscrew the two bolts, one below the pillion footrest, and the other on an extension of the hydraulic cylinder bracket. Now unscrew the eight exhaust pipe to cylinder head nuts (two on each pipe), and remove the complete exhaust with clamps and split collars.
3 Slacken the exhaust pipe to junction box clamp bolts and remove the two separate exhaust pipes, with gaskets.
4 When replacing the exhaust system, install first the silencer and integral exhaust pipes. Then put the gaskets on the two separate pipes and insert them into the junction box. Replace the retaining rings and split collars on the exhaust pipes. Fit new gaskets in the exhaust ports. Bolt up the cylinder head joints before tightening the exhaust pipe to junction box clips.

4 Piston - oil control ring

1 The piston now has a three piece oil control ring, consisting of top and bottom rails, with a spacer between them.
2 When fitting the oil control ring, first install the spacer, followed by the top and bottom rails. The spacer and rail gaps must be staggered 20 - 30 mm (0.8 - 1.2 inch).
3 The top and second compression rings must not be interchanged, install with their markings upwards. All ring gaps should be spaced at 120° intervals, but with none above the gudgeon pin bosses, nor exactly at the back or the front of the piston.
4 The stated ring gap of the oil control ring, is that of the spacer.

5 Clutch - arrangement of plates

1 The clutch assembly has been simplified. There is a strengtheing band round the clutch drum, and the outer plain plate is now integral with the clutch centre. Take careful note of the order of plates when dismantling, and replace them in the same order.
2 The outer friction plate (first to be assembled on the clutch drum) is slightly smaller than the others, and has tangential slots in the friction material. The slots must travel in the direction of rotation (clockwise), see Fig. 7.2.
3 There is now an external circlip behind the splined washer on the clutch shaft.

6 Carburettor - setting

1 The carburettor settings have been changed to suit the new exhaust system, and are as follows:

Choke diameter	28 mm
Main jet	105
Air jet	120
Slow jet	40
Air screw opening	1 turn $\pm$ 3/8
Cutaway number	2.5
Valve seat diameter	2 mm
Fuel level	26 mm (0.866 inch)
Jet needle setting	Third notch

7 Air filter and engine breather filter - cleaning and renewal

1 The air cleaner is removed for servicing or renewal by unscrewing the two bolts in the top of the filter casing, alongside the fixing brackets. Remove the lower half, with the filter.
2 The engine breather filter is fitted below the battery housing. Gases from the engine are passed through a micro-mesh filter, permitting oil and water to drain away, and the filtered gas is reintroduced into the carburettor intake. In this way any unburnt hydrocarbons are recycled through the engine.
3 Remove the left-hand side frame cover. The filter is secured by one hexagonal bolt to the base of the battery carrier. Remove this bolt, pull off the pipes and take out the filter housing. Take off the filter top, secured by two screws. Extract the gasket, spring, filter element retainer and filter. Wash the filter element (but not in petrol or other low flash point solvent), dry and replace. Make sure that the breather filter drain tube is clear; this tube goes to the bottom of the filter. The top hose goes to the top of the air filter, and the centre one to the engine breather. The lower end of all tubes, including the carburettor drain tubes, pass through a clip on the centre stand pivot clamp bolt.

8 Fuel filter and breather - cleaning

1 The fuel filter fitted to the CB 750F is no longer situated in a bowl below the tap. To clean the latest filter, it is necessary to drain the fuel tank, and remove the tap.
2 After draining the tank, unscrew the union nut above the fuel tap, and remove the tap and filter. Pull the filter out of the tap, and wash off any sediment. The base of the filter has a flat, which has to be located against a flat in the tap body. Do not omit the rubber washer over the filter gauge. The union nut is threaded right-hand for the tap. The straight portion of the nut goes adjacent to the tank. Offer up the tap to the tank boss, and tighten the union nut.
3 The tank incorporates a lockable cover for the filler, and a breather tube. The breather tube should be checked to make sure it is not blocked or cracked. Oil lightly the barrel of the lock, but **not** the internal mechanism. Should the lock have to be removed, it is held in place by a spring clip.
4 Take great care, when draining or refilling the petrol tank, to avoid fire or explosion.

9 Front forks - damper unit

1 The front forks can be removed as described for the CB 750 model, omitting the section on upper fork strouds; but the stanchions should be dismantled as follows:
2 **Note** that the stanchion filler cap should be slackened, but **not** removed, while the stanchion is still clamped in the yokes.
3 Remove the fork drain screw and pump the fork to eject all oil. Pull off the fork gaiter. Hold the lower fork leg in a vice with soft jaws; unscrew the hexagon socket cap screw in the bottom of the fork leg, **before** removing the stanchion filler cap; this prevents the damper from turning.
4 Now remove the filler cap with 'O' ring seal. Withdraw the fork spring. The spring is progressively wound, that is the coils are closer together at one end. The difference is slight, and note should be taken of the correct orientation when removing the spring.
5 Pull the stanchion and damper assembly from the lower fork leg. Push the damper out through the top end of the stanchion. Note the small rebound spring under the damper piston. Remove the wire circlip, and prize out the oil seal from the top of the slider.
6 Since the fork is no longer bushed, the bore of the lower fork leg should be examined for scores or wear. If there is any damage, the only action is to renew the lower fork leg. Also check the damper piston ring for wear or damage. Compress the ring carefully when re-inserting the damper into the stanchion. Use the fork spring to push the damper to the bottom, and replace the stanchion filler cap to prevent the damper from turning when tightening its fixing screw. Use thread lock compound on this screw, and renew the copper washer.
7 The headlamp is mounted in a frame, bolted to the upper fork yoke and fitted into bushes in the lower yoke.

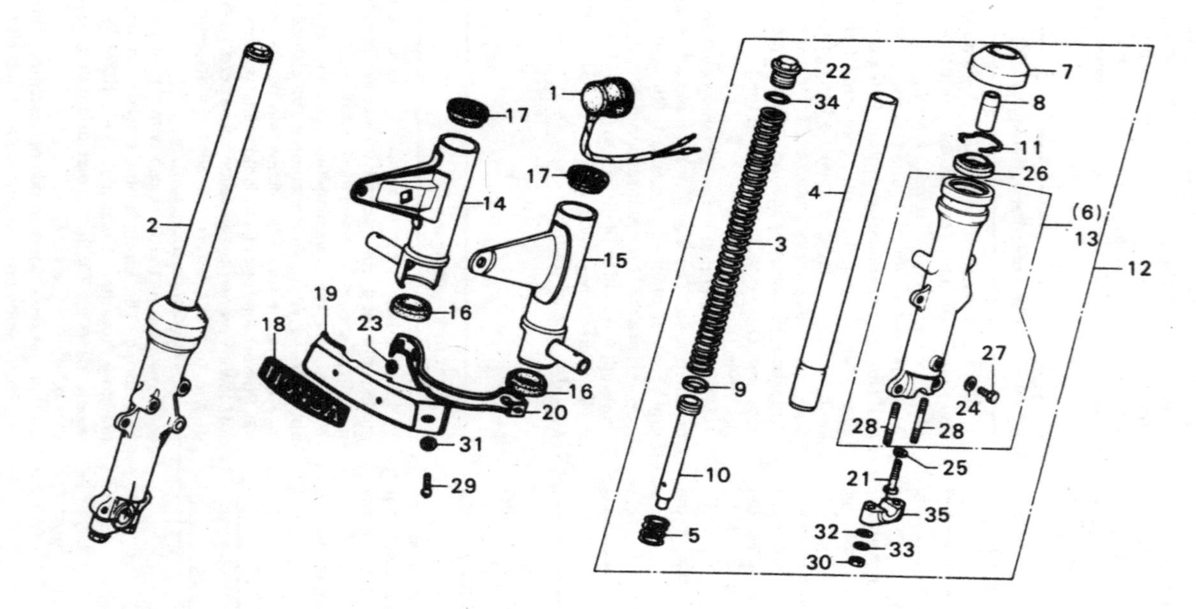

Fig. 7.3. Front forks

1	*Audible indicator warning unit*	*10*	*Damper rod - 2 off*	*19*	*Emblem fixing plate (K7 and A)*	*28*	*Stud - 4 off*
2	*Right-hand front fork assembly*	*11*	*Oil seal retainer ring - 2 off*	*20*	*Fixing plate stay (K7 and A)*	*29*	*Screw - 2 off (K7 and A)*
3	*Fork spring - 2 off*	*12*	*Front fork assembly complete - 2 off*	*21*	*Allen head screw - 2 off*	*30*	*Nut - 4 off*
4	*Fork stanchion - 2 off*	*13*	*Lower leg - 2 off*	*22*	*Fork top bolt - 2 off*	*31*	*Washer - 2 off (K7 and A)*
5	*Rebound spring - 2 off*	*14*	*Right-hand fork shroud*	*23*	*Nut - 2 off (K7 and A)*	*32*	*Washer - 4 off*
6	*Lower leg - 2 off*	*15*	*Left-hand fork shroud*	*24*	*Drain plug washer - 2 off*	*33*	*Spring washer - 4 off*
7	*Dust seal - 2 off*	*16*	*Lower shroud seating - 2 off*	*25*	*Washer - 2 off*	*34*	*O-ring - 2 off*
8	*Damper rod seat - 2 off*	*17*	*Upper shroud seating - 2 off*	*26*	*Oil seal - 2 off*	*35*	*Fork spindle clamp - 2 off*
9	*Piston ring - 2 off*	*18*	*Honda emblem (K7 and A)*	*27*	*Drain plug - 2 off*		

10 Swinging arm - renovation

1 The swinging arm may be removed as described for the CB 750 model except of course that rear wheel removal is as in Section 12 of this Chapter. In addition, detach the torque arm from the brake caliper by removing the split cotter pin and nut and bolt. Hook the caliper out of the way, still attached to the brake hose.
2 With the swinging arm removed from the frame, pull off the dust caps at each end of the pivot. These have integral seals. The long centre bush will push out, possibly after a gently tap. A flanged bush is pressed into each end of the pivot. Service the swinging arm as described for the CB 750.
3 The grease nipple on the CB 750F model is beneath the swinging arm pivot tube.

11 Prop stand - examination and renovation

1 This section applies to USA models only.
2 Check the rubber 'trip' on the end of the stand for wear or damage. No part should be worn below the moulded line on the rubber. Check the operation of the stand as follows: Place the machine on its centre stand, and put the prop stand down. Using a spring balance attached to the extreme end of the stand, measure the force required to retract the stand. If this force exceeds 2-3 kg (4.4 - 6.6 lb), check that the stand pivot bolt is not overtight, or in need of lubrication.
3 To renew the rubber 'trip', unscrew the bolt. Make sure the sleeve is installed in the fixing hole of the new trip. Fit the trip with the arrow facing outwards; the block should be marked 'OVER 260 LBS ONLY'.

12 Rear wheel - removal

1 The rear forks ends are now closed so that the rear wheel spindle must be withdrawn to remove the wheel.
2 Slacken the chain adjusters, and unhook the chain from the rear sprocket. This will be easier if the chainguard is removed, by unscrewing the two fixing bolts. Remove the split cotter pin on the right-hand side, and unscrew the castellated nut. Withdraw the rear wheel spindle using a tommy bar, being ready to catch the left-hand side spacer, and the chain adjusters.
3 The brake caliper may be left attached to the torque arm. While the wheel is removed **do not** operate the brake, as this will cause loss of fluid. It is a good idea to put a piece of wood between the pads, to prevent accidents.
4 After replacing the wheel, the rear chain must be re-adjusted.

13 Rear disc brake - examination

1 Remove the cap and diaphragm from the fluid reservoir, and check that fluid level is up to the maximum mark. Inspect all of the hydraulic system for leaks. Check the flexible hose for cracks or damage, and renew immediately if any are apparent.
2 Pad wear can be checked without removing the caliper. If one or both of the pads is worn, so that the tabs (painted red) come up to the central red mark on the caliper, both must be renewed.
3 The brake pedal height may be adjusted. The clearance between the brake pedal and the underside of the footrest should not be less than 5 mm (0.2 inches). Remove the split cotter pin on the master cylinder linkage, and take out the clevis pin. Slacken the locknut and turn the clevis fork, holding the hexagon above the locknut with a spanner to prevent it turning.
4 If the brake feels spongy, and there are no leaks, the system needs to be bled. Follow the procedure described in Chapter 5. Do not allow fluid to get on the tyre, or painted parts.
5 The brake disc should be examined for wear or distortion. The surface runout, measured with a dial test indicator should not exceed 0.3 mm (0.012 inch).

10.3. Swinging arm grease nipple

14 Rear disc brake - renovation

1 When the pads are worn to the red mark on the caliper, both pads must be renewed. Unscrew the hexagonal screw securing the pad cover to the caliper, and remove the cover. Press down on the pad retaining spring and pull out the retaining pin at the free end. Then pull out the other pin, and remove the spring and pair of pads. Assemble the new pads and spring and replace. Note that the spring engages in the groove in the pad retaining

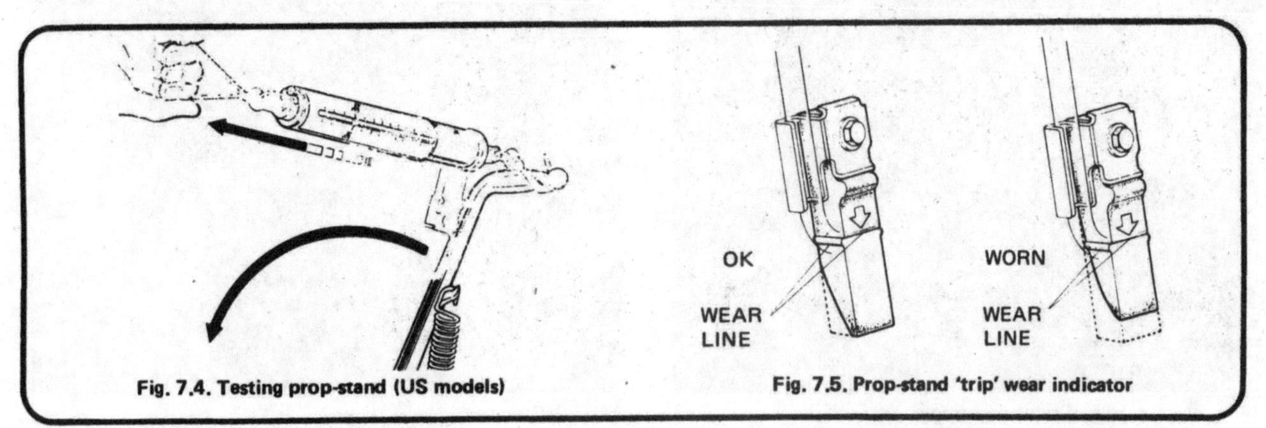

Fig. 7.4. Testing prop-stand (US models)

Fig. 7.5. Prop-stand 'trip' wear indicator

13.1. Check rear brake fluid level

13.2 Brake pad wear indicator tabs

13.3. Rear brake master cylinder linkage and adjuster

13.4. Rear brake bleed nipple

14.1. Rear brake pad retaining spring

14.3. Caliper torque arm fixing

pins, the chamfered end of the pins goes in first. Do not get grease or oil or brake fluid on the pad surfaces.
2 Before renewing the hydraulic hose, the fluid must be drained. Attach a tube to the bleed nipple, open the nipple, and pump the brake pedal until all fluid is expelled. Remove the rear wheel. Undo the banjo bolts at each end of the hydraulic hose, and remove the hose. There is a soft washer on each side of the banjo union. The hose passes through a clip on the swinging arm. Fill and bleed the system after replacing all parts.
3 If the caliper develops a leak, ie; fluid escapes around the piston, or if a piston siezes ie; the brake stays on or won't go on; the caliper must be removed for attention. This is best entrusted to an agent, if a new unit is not to be fitted. To remove the caliper, detach the torque arm after removing the split cotter pin and bolt. Partly withdraw the rear wheel spindle, and take out the caliper. Detach the hydraulic hose.
4 If the master cylinder requires attention, it too must be removed from the machine. Detach the hydraulic hose as previously described. Remove the split cotter pin and clevis pin from the brake rod. Unscrew the two cylinder fixing bolts, at the top of the cast aluminium bracket, and remove the cylinder. If the cylinder is not going to be renewed, overhaul should be entrusted to a dealer. Earlier models have a detachable reservoir.

15 Instrument heads - removal

1 The speedometer and rev-counter are now fixed to a bracket on the top yoke by two acorn nuts. To remove the instrument heads for overhaul, or to replace lamps, unscrew the nuts and remove the washers, bushes and sleeves. Unscrew the drive cable knurled nuts, and lift the instruments clear of the housing,

16 Electrical system

1 The main switch of the CB 750F has been moved to a more convenient positon, on the top fork yoke. Here, on USA models, it also locks the steering head. Electrical connections are made by a socket to an integral plug on the switch.
2 The three fuses now fitted, along with two spares, are in a holder beneath the left-hand side frame cover. The fuseholder cover can only be put on one way up. The three fuses are plainly marked, two 7 amp for head and taillamp respectively, and 15 amp for the main circuit.
3 The clutch operated starter switch is fitted into the clutch lever pivot. If a switch fault is suspected it may be checked for continuity only when the clutch is **disengaged**. The green and green with red switch wires are in the headlamp shell. The cut-out operates in conjunction with the neutral indicator switch and a logic diode. The plug-in diode is situated in a holder in the electrical compartment. Check the diode for continuity **only** in the direction shown by the arrow on the case (ie; positive lead to the 'blunt' end). Do not use the megohm (10^6 ohms) range for this test, or the diode may be damaged.
4 On American models, the headlamp comes on when the main switch is turned on. Consequently there is no lights switch on the handlebar. Also on USA models, the front indicators light up when on main beam. This is achieved with double filament lamps, as in the rear light. The lamps have offset pins, and are not interchangeable with the rear indicator lamps.
5 Also new, is the central warning lamp console, housing the five warning lamps. The lens cover is retained by three self-tapping screws. After unscrewing these, the cover may be removed. The lamps are bayonet fitting and can be quite tight. Do not overtighten the self-tapping screws or the threads may strip. The lamps are the same as those in the instrument heads.
6 A buzzer is fitted as an audible warning when the flashing indicators are on.

16.2. The fuses are clearly marked

16.3. The starter cut out diode plugs in

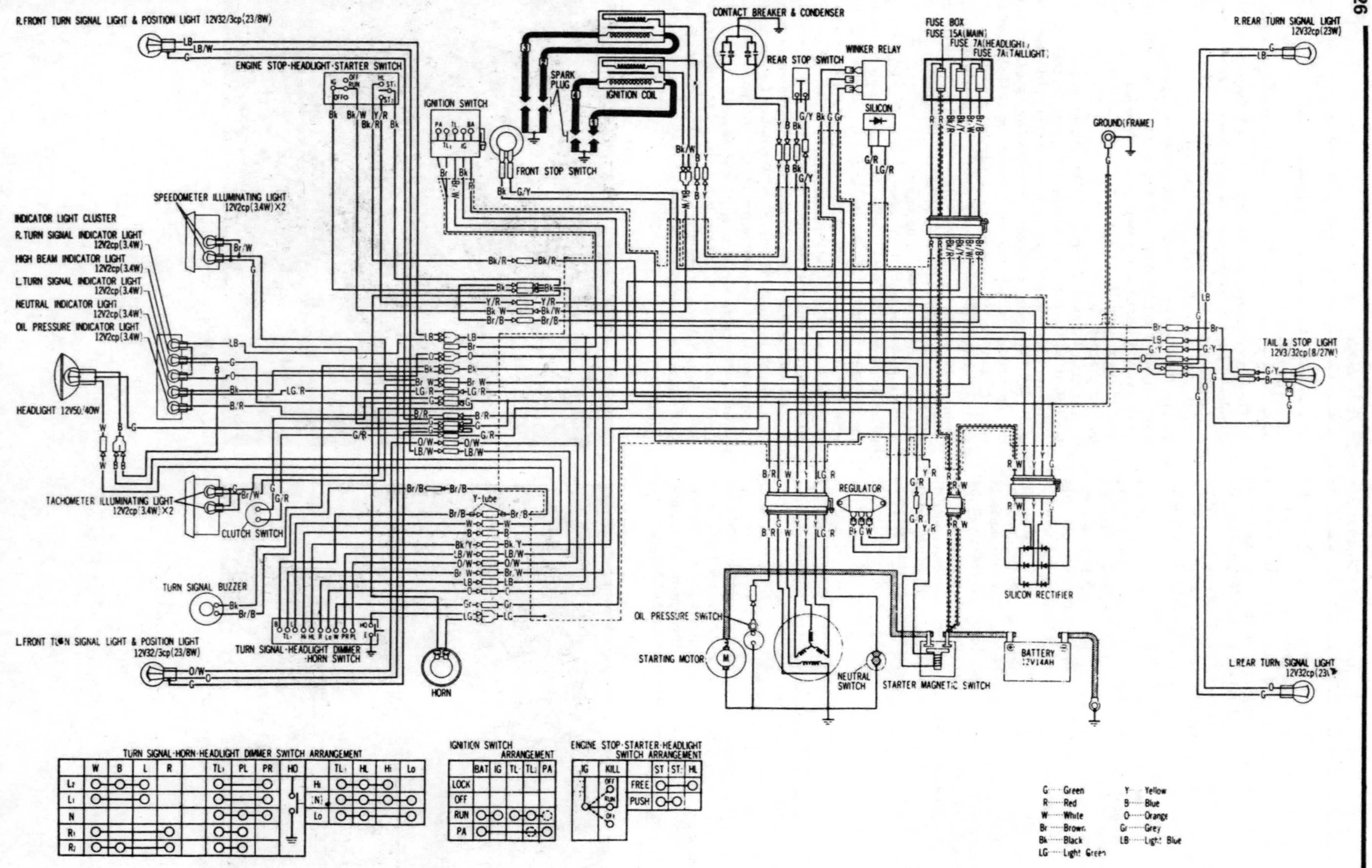

R.FRONT TURN SIGNAL LIGHT & POSITION LIGHT 12V32/3cp(23/8W)
ENGINE STOP·HEADLIGHT·STARTER SWITCH
IGNITION SWITCH
SPARK PLUG
IGNITION COIL
FRONT STOP SWITCH
CONTACT BREAKER & CONDENSER
REAR STOP SWITCH
WINKER RELAY
SILICON
FUSE BOX
FUSE 15A(MAIN)
FUSE 7A(HEADLIGHT)
FUSE 7A(TAILLIGHT)
R.REAR TURN SIGNAL LIGHT 12V32cp(23W)
GROUND(FRAME)
SPEEDOMETER ILLUMINATING LIGHT 12V2cp(3.4W)×2
INDICATOR LIGHT CLUSTER
R.TURN SIGNAL INDICATOR LIGHT 12V2cp(3.4W)
HIGH BEAM INDICATOR LIGHT 12V2cp(3.4W)
L.TURN SIGNAL INDICATOR LIGHT 12V2cp(3.4W)
NEUTRAL INDICATOR LIGHT 12V2cp(3.4W)
OIL PRESSURE INDICATOR LIGHT 12V2cp(3.4W)
HEADLIGHT 12V50/40W
TAIL & STOP LIGHT 12V3/32cp(8/27W)
TACHOMETER ILLUMINATING LIGHT 12V2cp(3.4W)×2
CLUTCH SWITCH
Y-tube
REGULATOR
TURN SIGNAL BUZZER
SILICON RECTIFIER
OIL PRESSURE SWITCH
STARTING MOTOR
NEUTRAL SWITCH
STARTER MAGNETIC SWITCH
BATTERY 12V14AH
L.FRONT TURN SIGNAL LIGHT & POSITION LIGHT 12V32/3cp(23/8W)
TURN SIGNAL·HEADLIGHT DIMMER·HORN SWITCH
HORN
L.REAR TURN SIGNAL LIGHT 12V32cp(23W)

TURN SIGNAL·HORN·HEADLIGHT DIMMER SWITCH ARRANGEMENT

IGNITION SWITCH ARRANGEMENT

	BAT	IG	TL1	TL2	PA
LOCK					
OFF					
RUN					
PA					

ENGINE STOP·STARTER·HEADLIGHT SWITCH ARRANGEMENT

G····Green
R····Red
W····White
Br····Brown
Bk····Black
LG····Light Green
Y····Yellow
B····Blue
O····Orange
Gr····Grey
LB····Light Blue

Fig. 7.6. Circuit diagram, CB750F (US model) *For UK diagram see page 150

1978 Honda CB 750 K7

Chapter 8 The CB750F2, K7 and A models

Contents

Specifications

Except where entered below, dimensions and specifications for the K7 and F2 are as given for the earlier models at the beginning of each Chapter. **Note** that 750A Specifications are given after the complete list of K7 and F2 Specifications.

Specifications relating to Chapter 1

Engine

	F2	K7
Compression ratio	9.0 : 1	9.2 : 1
Cylinder compression	142–170 psi (10.0–12.0 kg/cm²)	

Cylinder head

	F2	K7
Valve guide internal diameter (both valves)	6.60–6.62 mm (0.2598–0.2606 in)	
Valve spring free length		
Inner springs	40.30 mm (1.5866 in)	
Limit of wear	39.14 mm (1.5409 in)	
Outer springs	43.05 mm (1.6949 in)	
Limit of wear	41.74 mm (1.6433 in)	
Stem to guide clearance		
Inlet valves	0.10 mm (0.0039 in) maximum limit	
Exhaust valves	0.13 mm (0.0051 in) maximum limit	
Valve clearance (cold)		
Inlet valves	0.05 mm (0.0020 in)	
Exhaust valves	0.08 mm (0.0031 in)	
Valve timing		
Inlet opens	5° BTDC	0° BTDC
Inlet closes	40° ABDC	40° ABDC
Exhaust opens	40° BBDC	40° BBDC
Exhaust closes	5° ATDC	0° ATDC
Cylinder head nut torque setting	2.0–2.5 kg m (14.5–18.1 lb ft)	

Camshaft

Cam height	
Inlet	36.00 mm (1.4173 in)
Exhaust	35.50 mm (1.3976 in)
Camshaft cap nut/bolt torque setting	0.6–1.1 kg m (4.3–8.0 lb ft)

Piston rings

End gap (compression)	0.2–0.4 mm (0.008–0.016 in)
Limit of wear	0.7 mm (0.028 in)
End gap (second compression)	0.2–0.4 mm (0.008–0.016 in)
Limit of wear	0.7 mm (0.028 in)
End gap (oil control)	0.3–0.9 mm (0.012–0.035 in)
Limit of wear	1.2 mm (0.047 in)
Side clearance (compression rings)	0.8 mm (0.0071 in) maximum
Side clearance (oil control ring)	0.165 mm (0.0065 in) maximum

Clutch

Clutch springs - free length	39.70 mm (1.5630 in)	31.94 mm (1.2575 in)
Limit of wear	36.80 mm (1.4488 in)	30.50 mm (1.2008 in)

Gearbox

Type	Five-speed constant-mesh
Gear ratios:	
1st gear	2.500 : 1
2nd gear	1.708 : 1
3rd gear	1.333 : 1
4th gear	1.133 : 1
5th gear	0.969 : 1
Primary drive ratio	1.708 : 1

Specifications relating to Chapter 2

Fuel tank capacity	Overall	Reserve
750 F2	17.5 lit (4.8 US/3.9 Imp gal)	4.5 lit (1.2 US/1.0 Imp gal)
750 K7	19.5 lit (5.1 US/4.3 Imp gal)	4.0 lit (1.1 US/0.9 Imp gal)

Carburettors	F2	K7
Make	Keihin	Keihin
Type	Piston valve	Piston valve
Choke size	28 mm	28 mm
Main jet	105	115
Slow running jet	35	35
Main air jet	120	150
Pilot air jet	150	150
Jet needle setting	F2051F–2	F2D51E–1
Pilot jet setting	1¾ turns out	1½ turns out
Float level	14.5 mm (0.5709 in)	12.5 mm (0.4921 in)
Idle speed	1000 rpm ± 100 rpm	
Oil tank capacity	3.5 lit (7.4 US/6.1 Imp pints)	

Specifications relating to Chapter 3

Ignition timing

On 'F' mark	10° BTDC @ 1000 rpm
Full advance	35° BTDC @ 2500 rpm

Spark plugs

Make	NGK	Nippon Denso
Type		
UK	DR8ES–L	X24ESR–U
USA	D8EA	X24ES–U

Specifications relating to Chapter 4

Front forks

Type	Telescopic, hydraulically damped
Travel	141.5 mm (5.6 in)
Fork spring free length	504.3 mm (19.854 in)
Fork spring service limit	495.0 mm (19.488 in)
Fork piston - outer diameter (max)	34.9 mm (1.374 in)
Lower fork leg - internal diameter (max)	35.25 mm (1.387 in)
Fork oil capacity (per leg)	145–155 cc (4.8–5.2 oz)
Refill after draining	135–145 cc (4.5–4.8 oz)

Rear suspension

Type	Swinging arm, supported on two hydraulically damped suspension units
Travel:	
F2	86.3 mm (3.4 in)
K7	101.6 mm (4.0 in)
Suspension units	
Free length of spring	232.9 mm (9.1693 in)
Service limit	228.9 mm (9.0118 in)

Specifications relating to Chapter 5

	F2	K7
Tyres		
Front	3.25H 19 (4PR)	3.50H 19 (4PR)
Rear	4.00H 18 (4PR)	4.50H 17A (4PR)
Tyre pressures		
Front	28/32* psi (2.0/2.25 kg cm^2)	28 psi (2.0 kg cm^2)
Rear	28/40* psi (2.0/2.8 kg cm^2)	32/40* psi (2.25/2.8 kg cm^2)

*Higher pressures for sustained high speeds and/or carrying a passenger regularly

	F2	K7
Brakes		
Front	Hydraulically operated twin discs	Hydraulically operated single disc
Rear	Hydraulically operated single disc	Single leading shoe drum

Specifications relating to Chapter 6

	F2	K7
Bulbs		
Headlamp	60/55w 50/40w sealed beam (USA)	60/55w
Tail/stop	5/21w 3/32w (USA)	5/21w 3/32w (USA)
Instrument lights	3.4w	3.4w
Pilot light (not fitted to USA models)	4w	4w

High beam indicator	3.4w
Neutral indicator	3.4w
Oil pressure warning	3.4w
Indicator repeater lights	3.4w
Flashing indicators	21w 8/23w (USA)

Starter motor	
Brush length (minimum)	7 mm (0.2756 in)

Dimensions and weights

	F2	K7
Overall length	2210 mm 87.0 in	2280 mm 89.8 in
Overall width	860 mm 33.9 in	880 mm 34.6 in
Overall height	1185 mm 46.7 in	1185 mm 46.7 in
Wheelbase	1480 mm 58.3 in	1495 mm 58.9 in
Seat height	830 mm 32.7 in	810 mm 31.9 in
Ground clearance	135 mm 5.3 in	150 mm 5.9 in
Weight (dry)	232.5 kg 512.6 lb	231 kg 508 lb

Except where shown below, specifications for the CB750A model are as given for the original CB750.

Specifications relating to Chapter 1

Engine	
Compression ratio	8.6 : 1
Valve timing	
Inlet opens	5° ATDC
Inlet closes	30° ABDC
Exhaust opens	40° BBDC
Exhaust closes	5° BTDC
Valve clearance	
Inlet valves	0.05 mm (0.002 in)
Exhaust valves	0.08 mm (0.003 in)
Cylinder head nut torque setting	2.0–2.5 kg m (14.5–18.1 lb ft)
Camshaft	
Cam height	
Inlet	35.314 mm (1.3903 in)
Exhaust	34.893 mm (1.3737 in)

Piston rings

End gap (compression)	0.2–0.4 mm (0.008–0.016 in)
Limit of wear	0.7 mm (0.028 in)
End gap (second compression)	0.2–0.4 mm (0.008–0.016 in)
Limit of wear	0.7 mm (0.028 in)
End gap (oil control)	0.3–0.9 mm (0.012–0.035 in)
Limit of wear	1.2 mm (0.047 in)
Side clearance (oil control)	0.165 mm (0.0065 in) maximum

Clutch

Clutch return spring free length	39.7 mm (1.56 in)
Clutch disc thickness	1.9 mm (0.075 in) minimum
Clutch plate thickness	1.9 mm (0.075 in) minimum
Clutch end plate thickness: standard	
(1) 1.8 mm (0.071 in)	1.6 mm (0.063 in) minimum
(2) 2.1 mm (0.083 in)	1.9 mm (0.075 in) minimum
(3) 2.4 mm (0.094 in)	2.2 mm (0.087 in) minimum
(4) 2.7 mm (0.106 in)	2.5 mm (0.098 in) minimum
(No mark) 3.0 mm (0.118 in)	2.8 mm (0.110 in) minimum
(6) 3.3 mm (0.130 in)	3.1 mm (0.122 in) minimum

Gearbox

Type	Two-speed with torque converter
Primary reduction ratio	1.351 : 1 (1.349 : 1 '77 and '78 models)
Gear ratios (I)	2.263 : 1 N – L – D or N – 1 – 2
(II)	1.520 : 1
Final reduction ratio	2.824 : 1 (2.800 : 1 '77 and '78 models)

Specifications relating to Chapter 2

Fuel tank capacity

Overall	19.5 lit (5.1 US/4.2 Imp gal)
Reserve	4.0 lit (1.1 US/0.9 Imp gal)

Carburettors

	'76 model	'77 model	'78 model
Main jet	102	108	108
Slow running jet	38	38	38
Main air jet	150	200	200
Pilot air jet	150	150	150
Jet needle setting	3rd notch	3rd notch	
Pilot jet setting	1¼ turns out	1 turn out	1.1/8 turns out
Float level	14.5 mm (0.571 in)	12.5 mm (0.492 in)	
Idle speed		950 ± 100 rpm in N	
Oil sump capacity		5.5 lit (5.8 US/4.8 Imp qt)	
After draining		4.0 lit (4.2 US/3.5 Imp qt)	

	'76 model	'77 model	'78 model
Oil pressure		50–64 psi @ 3000 rpm	
Transmission oil pressure		57–114 psi @ 1500 rpm	
Minimum permissible pressure		43 psi @ 1500 rpm	

Oil pump

Inner rotor/outer rotor clearance	0.20 mm (0.0079 in) max
Outer rotor/body clearance	0.35 mm (0.0138 in) max
Rotors to body clearance (endwise)	0.08 mm (0.0031 in) max

Specifications relating to Chapter 3

Ignition timing

On 'F' mark	10° BTDC @ idle speed
Full advance	33° – 36° BTDC @ 2500 rpm

Spark plugs

Make	NGK	Nippon Denso
Type:		
1976 and 1977 models	D8ES–L	X24ES
1978 model	D8EA	X24ES–U

Specifications relating to Chapter 4

Front forks

Type	Telescopic, hydraulically damped
Travel	141.5 mm (5.6 in)
Fork springs - free length	504.3 mm (19.85 in)
Service limit	495.0 mm (19.5 in)

Fork piston - outer diameter (max)	34.9 mm (1.374 in)
Lower fork leg - internal diameter (max)	35.25 mm (1.388 in)
Fork oil capacity	145–155 cc (4.9–5.3 oz)
Refill after draining	135–145 cc (4.6–4.9 oz)
Rear suspension	
Type	Refer to F2 and K7
Travel	91.5 mm (3.6 in)
Suspension units:	
Free length of spring	232.7 mm (9.16 in)
Service limit	220.0 mm (8.66 in)

Specifications relating to Chapter 5

Tyres	
Front	3.50 H 19 (4PR)
Rear	4.50 H 17A (4PR)
Tyre pressures	
Front:	
1976 and 1977 models	25/28* psi (1.75/2.0 kg cm^2)
1978 model	28/28 psi (2.0/2.0 kg cm^2)
Rear:	
1976 and 1977 models	32/36* psi (2.25/2.5 kg cm^2)
1978 model	28/36* psi (2.0/2.5 kg cm^2)

*Higher pressures for sustained high speeds and/or regularly carrying a passenger.

Specifications relating to Chapter 6

Battery	
Capacity	20 Ah
Bulbs	
Headlamp	50/40w
Tail/stop	8/27w (3/32 cp)
Turn signal lights	23w (32 cp)
Speedometer light	3.4w x 2 (2 cp x 2)
Change indicator light	3.4w x 3 (2 cp x 3)
Turn signal indicator light	3.4w x 2 (2 cp x 2)
High beam indicator light	3.4w (2 cp)
Gear position light	8w x 2 (3 cp x 2)

Dimensions and weights

	CB750A	
Overall length	2250 mm (88.6 in)	
Overall width	865 mm (34.1 in)	
Overall height	1185 mm (46.7 in)	1190 mm (46.9 in)*
Wheelbase	1480 mm (58.3 in)	
Seat height	820 mm (32.3 in)	810 mm (31.9 in)*
Ground clearance	135 mm (5.3 in)	140 mm (5.5 in)*
Weight (dry)	241 kg (531 lbs)	245 kg (540 lbs)*

*Relates to the 1977 and 1978 models.

1 General description - CB750 F2 and K7 models

The CB750 K7 and F2 were introduced in May and June of 1977 respectively. They are largely similar to their forerunners, the K6 and F1, although both of the newer models are substantially modified both in specification and styling.

The design features incorporated in the K7 which differ materially from its immediate predecessor are as follows.

a) A larger (4.2 Imp gal) petrol tank.
b) Redesigned seamless silencers.
c) Smaller (17 inch diameter), rear wheel, with a wider (4.50 inch section), tyre.
d) Heavy-duty ¾ inch pitch drive chain with sealed-in lubricant.
e) A quartz halogen headlamp is now fitted (not to US models) as are larger indicator lenses and a larger tail light.
f) The choke knob is repositioned on the handlebar clamp and the ignition switch is re-sited between the base of the instrument heads (US models already had the latter modification, and now have an integral ignition switch/steering lock).

In addition to these changes, the engine in the K7 is the same as that fitted to the earlier F1 model with the addition of the latest type carburettors. These now feature an accelerator pump and enclosed linkages.

The modified design features of the F2 model also include points (d), (e) and (f), plus several others. Further, the engine fitted to this, the latest of the 'Super Sport' models, has had substantial modifications carried out in order to boost power output. The modifications are listed below.

a) There are larger valves fitted. Inlet valves are 34 mm, whilst exhaust valves are 31 mm. This is a 2 mm and 3 mm increase respectively over the F1.
b) The compression ratio is lowered from 9.2:1 to 9.0:1.
c) The combustion chamber and port shape have been improved, and the camshaft profile has been modified altering the cam timing.
d) Stiffer valve springs are fitted. They are rated at 202 pounds as opposed to 154 pounds on the F1.
e) New 28 mm carburettors, complete with an accelerator pump and enclosed linkages are fitted.
f) To cope with the performance increase there is a larger oil filter fitted, more cooling fin area in the sump and stiffer

clutch springs, uprated from 220 pounds (F1) to 242 pounds.

Besides these engine modifications, there are several other design and styling innovations and modifications to the F2.

g) Modified front fork internal components.
h) New aluminium castings carrying the rider's and pillion's footrests, and the rear brake lever on the right.
i) The silencer is of a different design.
j) There is a twin front disc brake set-up, complemented by a single disc brake on the rear wheel. The front disc calipers are now mounted behind the fork legs.
k) The wheels are of the Comstar variety, and are of a composite form, with an aluminium alloy rim and pressed steel spokes.
l) The electrical connections, which used to be mounted in a box on the front left-hand frame downtube of the F1, are now fitted inside the headlight shell.
m) Twin horns are now fitted.

Apart from these innovations and certain cosmetic changes, such as the matt black paint finish to the engine, front fork stanchions and rear shock absorber springs, the F2 is the same as its immediate predecessor.

The information contained in the preceding Chapters can be applied generally to the F2 and K7 models. Where there are fundamental procedural differences they will be dealt with in this Chapter.

2 Points to note - engine and gearbox (F2 only)

1 When replacing the cylinder barrel and/or the cylinder head, note that there are now fewer dowels. The number of the 12 mm dowels and their rubber sealing sleeves on the inner studs has been reduced from eight to four.

2 The K7 continues with the previous arrangement.

3 In the gearbox the only change to note is that the previous arrangement of separate bushes and thrust washers on the gearbox layshaft has given way to integral bush and thrust washer components.

3 Petrol tank - removal and replacement

1 To remove the petrol tank, turn off the petrol tap and detach the petrol pipe and protective spring (K7 only), after first releasing the wire retaining clip. Also detach the overflow pipes and their connecting joints which run under the bottom left edge of the tank.

4 Carburettors - removal

1 To enable the carburettors to be removed, the petrol tank must first be lifted clear.

2 The throttle and choke cables must be removed from the cable holder and then disconnected from the choke shaft levers.

3 If, when the clamps that hold the carburettors to the inlet stubs and air cleaner box are slackened, the carburettors still cannot be manoeuvred free, then the air cleaner case must be removed.

4 Remove the air cleaner lower case by loosening the two mounting bolts. With the carburettor clamps already slack, the two upper half case bolts can be removed. The two parts can then be lifted clear.

5 At this stage the carburettors can be left as a complete unit, still attached to their mounting plate.

5 Carburettors - dismantling, examination and reassembly

1 Before separation and dismantling of the carburettors is begun it should be remembered that it is essential to keep separate the identical components from each carburettor. To prevent inadvertent interchange store the components in separate, clearly marked containers, and where possible dismantle the carburettors one at a time.

2 Remove the rear stay from the carburettor assembly by unscrewing the four bolts. Unhook the throttle return spring and detach the mounting plate from the front of the carburettors. The eight screws which hold the mounting plate in position are often exceptionally tight and care should be taken not to damage the screw heads during initial loosening.

3 Detach the choke butterfly valve from each carburettor. Each valve is held to the transverse operating shaft by two screws which are secured by being peened over. If necessary file off the peened area before removing the screws.

4 Remove all four carburettor tops; each is retained by two screws. Access can now be made to the throttle operating bellcranks. Unscrew the screw or bolt that passes through the end of each bellcrank into the transverse throttle operating rod, and remove the screw from the throttle operating pulley boss after slackening the locknut.

5 Displace the two spring clips on each rubber accelerator pump fuel feed tube towards the centre of the tube. This will allow the tubes to slip off their unions more easily when the carburettors are separated.

6 Disconnect the accelerator pump return spring and release the end of the choke relief spring. The position of the relief spring and the amount it is tensioned should be noted to aid reassembly.

7 The carburettors can now be separated from the throttle operating rod. Slide off the two left-hand carburettors and displace the shared choke operating rod. Remove the far right-hand carburettor and withdraw the choke operating rod from the inner right-hand (No 3) carburettor. To enable the No 3 carburettor to be removed from the throttle rod the throttle stop arm, which lies outboard of the carburettor, must be displaced. The arm is secured by a roll pin which passes through the arm boss into the throttle rod. The roll pin must be drifted out using a parallel sided drift of suitable size, with the link arm supported on wooden blocks.

8 With all the carburettors separated from one another dismantling of the individual carburettors is now possible. For the reasons given in paragraph 1, dismantle the carburettors separately.

9 Withdraw the bellcrank/throttle valve assembly from the carburettor. If throttle valve or needle renewal is required these components may be detached by unscrewing the two small screws which pass through the link plate into the valve top.

10 To gain access to the various jets it is first necessary to remove the float chamber bowl by removing three screws. The float arm pin can then be pulled out allowing the float to be lifted out. The main jet can now be removed. Do NOT remove the slow running jet as it is a tight pressure fit.

11 The pilot air screw, jet needle, needle jet and float valve should be examined for hardened deposits, grooving, or any other damage and replaced if necessary. If damage to the float valve seat is found the complete carburettor body must be renewed. The main and slow running jets should be inspected for any evidence of clogging, and if there is any foreign matter present, it should be blown clear with compressed air. The use of a tyre pump makes an acceptable substitute if a source of compressed air is not available. Never use wire to remove deposits from carburettor jets or passageways; the finely metered bores are easily made oversize. Examine the throttle slide for wear. If it has become a very loose fit in the carburettor body carburettor adjustment problems will develop and slow running will become erratic. In extreme cases renewal of the valve alone

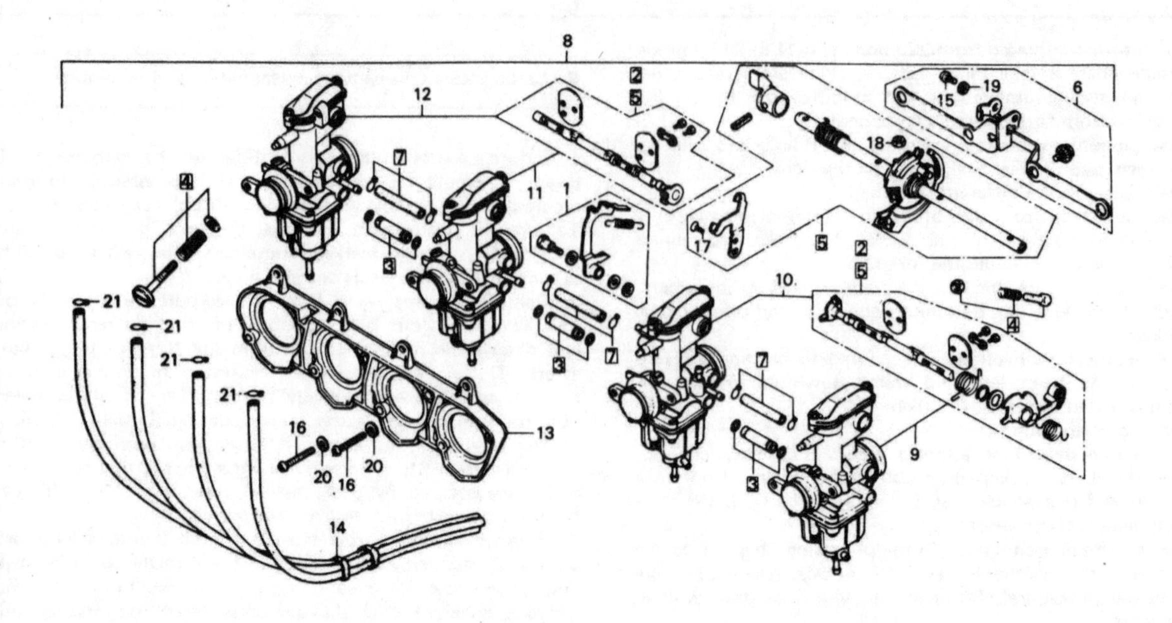

Fig. 8.1. Carburettor assembly and linkage - F2, K7 and A

1 *Pump link arm set*
2 *Choke link arm set - 2 off*
3 *Fuel transfer pipe set*
4 *Set screw*
5 *Throttle link set - 2 off*
6 *Rear stay set*
7 *Accelerator pump transfer pipe set*
8 *Carburettor assembly complete*
9 *Carburettor assembly No. 1*
10 *Carburettor assembly No. 2*
11 *Carburettor assembly No. 3*
12 *Carburettor assembly No. 4*
13 *Front support plate*
14 *Overflow tube assembly*
15 *Screw*
16 *Screw - 8 off*
17 *Screw*
18 *Nut*
19 *Spring washer*
20 *Spring washer - 8 off*
21 *Overflow tube spring clip - 4 off*

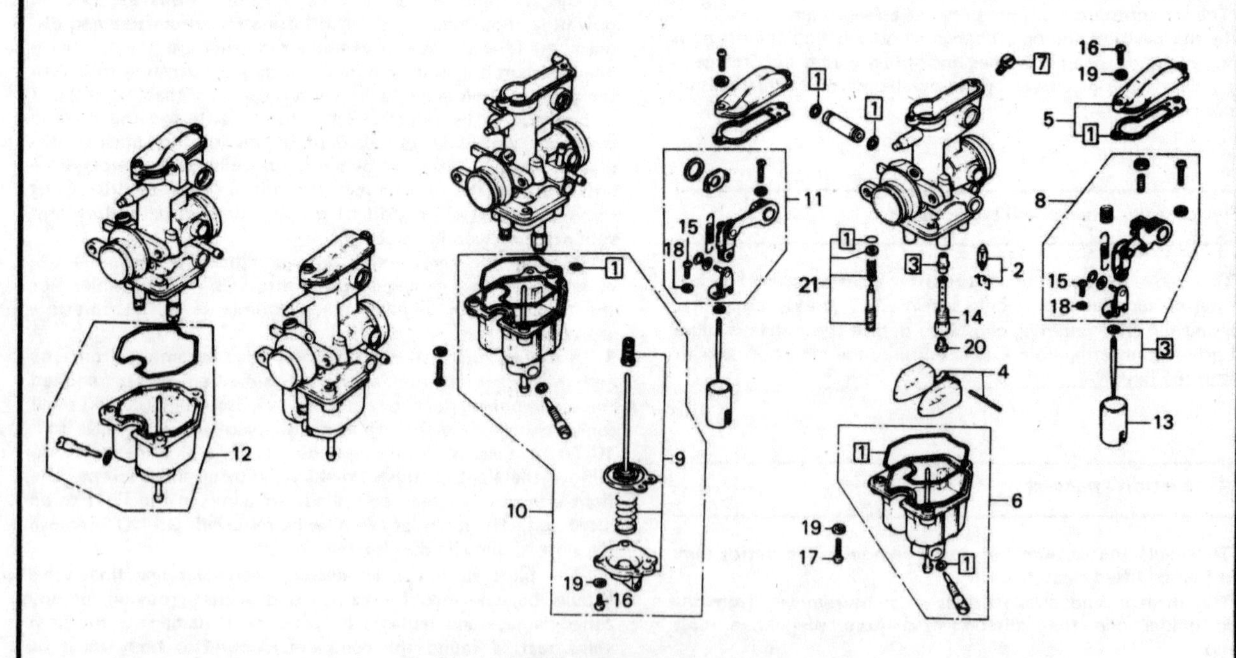

Fig. 8.2. Carburettors components - F2, K7 and A

1 *O-ring/gasket set - 4 off*
2 *Float valve set - 4 off*
3 *Jet needle set - 4 off*
4 *Float set - 4 off*
5 *Carburettor top cover set - 4 off*
6 *Float chamber set - 4 off*
7 *Blanking screw - 4 off*
8 *Link arm set - 3 off*
9 *Accelerator pump diaphragm set*
10 *Float chamber set (No. 2 carburettor)*
11 *Link arm set (No. 2 carburettor)*
12 *Float chamber set - 2 off*
13 *Throttle valve - 4 off*
14 *Jet needle holder - 4 off*
15 *Screw - 8 off*
16 *Screw - 11 off*
17 *Screw - 12 off*
18 *Spring washer - 8 off*
19 *Spring washer - 23 off*
20 *Main jet - 4 off*
21 *Pilot air screw assembly - 4 off*

will be insufficient and the complete carburettor must be renewed.
12 The accelerator pump, which is fitted to No 2 carburettor, is unlikely to give trouble. Examination of the pump diaphragm can be made after releasing the pump cover from the base of the float bowl, where it is secured by three screws. If the diaphragm has perished or is perforated it should be renewed.
13 Reassembly of the carburettors and installation of the throttle control and choke control rods should be carried out by reversing the dismantling procedure. Care should be taken to keep the working area and components scrupulously clean. In addition to ensuring that parts are fitted in their original position the following points should be noted.
14 Each throttle valve should be refitted to its bellcrank with the cutaway on the valve towards the choke valve, when installed in the carburettor body. Note that the bellcrank which is **not** fitted with an adjuster screw should be installed in the No 2 carburettor.
15 When refitting the choke valves on the control rods new screws should be fitted together with double tab washers. On original assembly the washers were not fitted, the screws being secured by peening or staking. If the correct washers are not available apply thread locking fluid to the screw threads before they are inserted. If the tab washers are used remember to bend up the ears of the washers to secure the hexagonal screw heads after final tightening.
16 It is recommended that a new roll pin is used when fitting the throttle stop arm to the throttle operating rod outboard of No 3 carburettor.
17 Do not fit the carburettor top to any but the No 2 carburettor until after final synchronisation has taken place as described in the following Section.

6 Synchronising the carburettors

1 To enable the fitting of the vacuum gauge set and subsequent testing for carburation balance it is not necessary to remove the air cleaner case.
2 If, when the gauges are connected, the engine is at its correct, normal running temperature, and at the correct idling speed, the readings are not the same on all four gauges, then the carburettors must be adjusted.
3 The carburettor tops of No's 1, 3 and 4 must be removed. This will allow access to the adjusting screw and locknut. The locknut should be loosened and the screw turned until the vacuum gauge readings become the same as on the base No 2 carburettor.
4 If it becomes necessary to adjust the accelerator pump assembly, this must be done with the carburettors removed from the engine. The pump rod to pump arm clearance, with the throttle throttle valve closed, should be measured. The clearance should be 0–0.2 mm (0.008 in). The pump arm to carburettor stay clearance, again with the throttle valve closed, should also be measured. The clearance here should be 9.5–10.5 mm (0.374–0.413 in). If it is necessary, the clearance can be adjusted by carefully bending the pump arm.

7 Carburettors - adjustment

1 Before adjusting the carburettors, a check should be made to ensure that the following settings are correct; contact breaker gap, ignition timing, valve clearances, spark plug gaps, crankcase oil level. It is also important that the engine is at normal running temperature.
2 The normal idling speed (1000 rpm) can be set by adjusting the stop screw. This is fitted at the front of the carburettor unit between carburettor No's 1 and 2. To increase the idling speed the throttle stop screw must be turned clockwise. Conversely decreasing the idling speed necessitates turning the screw anticlockwise.
3 If the correct idling speed cannot be obtained by this method the pilot air screw should be readjusted. The screw should be turned in until it lightly seats, and then backed out to the required setting. On the F2 that is 1¾ turns out, and on the K7, 1½ turns out. With the pilot air screw correctly set, readjust the idling speed by means of the stop screw.
4 The speed at which the engine idles when on choke, the fast idle speed, should be adjusted with the engine just started. The engine should idle at approximately 2750 rpm, with the choke operating knob fully extended. If the engine idles at more than 750 rpm below or above 2750 rpm with the choke knob in this position, adjustment is necessary.
5 To enable access to be made to the adjusting screw, situated between carburettors 2 and 3, the petrol tank must be removed (Section 3, this Chapter). Slacken the locknut on the adjuster screw and turn the screw inwards until it just contacts the choke operating cam. Push in the choke operating knob fully, and turn the adjuster screw in another 1½ to 2 turns. The fast idle speed should now be correct. Retighten the adjuster screw locknut and replace the petrol tank.

8 Footrests - removal (F2 model)

1 The rider's and pillion's footrests now bolt onto a single aluminium casting on each side of the machine.
2 On the left side there are three retaining bolts. Two serve to hold on the rider's and pillion's footrests and the third is a through bolt to hold the casting on to the frame, and act as the swinging arm pivot.
3 The casting on the right side also carries the rear brake actuating lever. This is fitted to a splined shaft which emerges out of the casting. To the rear of this shaft is a rectangular slot in which is fitted a rubber cap. Behind this cap is the adjustment for the rear brake operating lever. There are again two bolts, for both footrests, and the castellated nut on the end of the swinging arm pivot bolt.

9 Wheels - description

1 Both the CB750 F2 and K7 have received considerable attention to their wheels and tyres. The wheels on the F2 are of the Comstar variety. These consist of an aluminium alloy rim and hub, interconnected by pressed steel plate spokes. The components are riveted together to form a unit, the strength of which is potentially far greater than that of the traditional wire spoked wheel. The latter type of wheel is retained on the K7 albeit in differing rim size. Both the models have 19 in diameter front wheels, the F2 having a 3.25 in section tyre fitted, and the K7 a 3.50 in section. At the rear the F2 has an 18 in diameter wheel carrying a 4.00 in section tyre, whilst the K7 has a smaller diameter, 17 in, wheel, and a wider 4.50 in section tyre fitted.
2 The front wheel of the F2 carries a twin disc brake and the rear wheel a further single disc. The K7 has a single disc on the front wheel and a single leading shoe drum brake on the rear wheel.

10 Wheels - examination and renovation

1 The traditional wire spoked wheels as fitted to the K7 can be maintained and if need be, renovated, by reference to the existing information in Chapter 5 (Section 2).
2 The Comstar wheels of the F2, however, offer no such scope for adjustment or renovation. The wheels cannot be repaired if buckling or damage to the spokes has occurred. In the event of such damage, the wheel must be renewed as a complete unit. Do NOT attempt to tighten or loosen the bolts which hold the

spoke plates to the wheel hub. The nuts are tightened and secured by pins at the factory and should not be disturbed.
3 Using a fixed pointer, check the wheel rim for misalignment when it is rotated. Honda recommend the following maximum permissible limits for both wheels:

Radial run-out	2.0 mm (0.08 in)
Axial run-out	2.0 mm (0.08 in)

If the rims show more misalignment than this, then again, the whole wheel must be replaced.

11 Replacing brake pads

1 Remove the front wheel by following the procedure described in Chapter 4, Section 2.7.
2 The F2, with its new twin disc front brake set-up, has different calipers and pads to those of the K7. The caliper type, however, is similar to that fitted to the rear wheel of the F1. Refer to Chapter 7, Sections 13 and 14 for the brake pad renewal procedure. The procedure for K7 pad renewal is described below.
3 The front brake pads require replacement when the red line on top of the pads reaches the edges of the brake disc.
4 There are two caliper mounting bolts; these must be removed. The caliper, as a complete unit with the pads and shims still in place, can then be swung free. It is not necessary to disconnect the brake fluid hoses when removing the calipers although great care must be taken not to damage the hoses, as even small splits or cracks can lead to complete loss of braking.
5 It is also imperative that the brake lever is not operated during caliper removal and pad removal, as this will displace the piston and cause brake fluid seepage.
6 The two pads and the anti-chatter shim can now be removed.
7 Clean out any dust or dirt that has accumulated in the pad recesses and also the exposed end of the piston. A small soft brush should be used for this, NOT a wire brush.
8 When installing the new pads care should be taken to ensure that the arrow on the shims is facing upwards. It will be necessary to push the piston firmly inwards to allow the new pads to seat properly.
9 Sections 3, 5 and 7 apply to the type of caliper fitted to both the F2 and K7. There are, however, differences in the dismantling procedure of the single disc front brake on the K7.
10 The brake hose can again be left undisturbed at its junction on the caliper although it must be freed from its holding clip on the mudguard. There are again two bolts which, when removed, release the caliper. There is also a split pin to be removed. This goes through the rear of the caliper and locates in slots at the top of each pad. Do NOT omit to re-insert the split pin.
11 When inserting the two new pads apply a little silicon grease to the inserted faces of each before reassembly.
12 The caliper will now require readjustment. The lock nut on the fork lower stanchion should be loosened. Unscrew the caliper adjusting bolt fully. With the pads inserted and the brake hose refitted in its clip, the front wheel should be raised off the ground. Screw in the adjusting bolt until the inner pad, nearest the wheel, makes contact with the brake disc, and then screw it out ½ turn. Retighten the locknut.

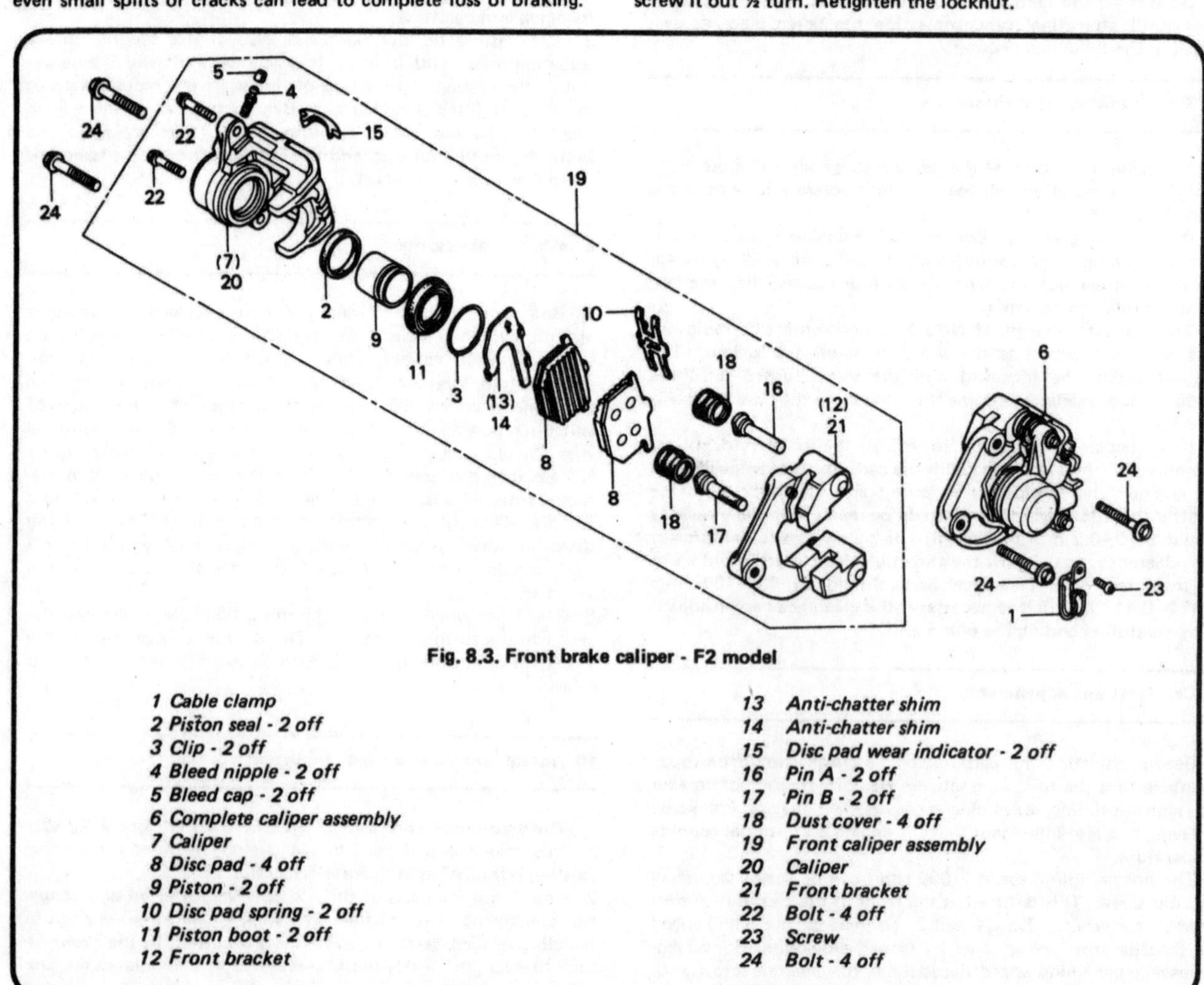

Fig. 8.3. Front brake caliper - F2 model

1 Cable clamp
2 Piston seal - 2 off
3 Clip - 2 off
4 Bleed nipple - 2 off
5 Bleed cap - 2 off
6 Complete caliper assembly
7 Caliper
8 Disc pad - 4 off
9 Piston - 2 off
10 Disc pad spring - 2 off
11 Piston boot - 2 off
12 Front bracket
13 Anti-chatter shim
14 Anti-chatter shim
15 Disc pad wear indicator - 2 off
16 Pin A - 2 off
17 Pin B - 2 off
18 Dust cover - 4 off
19 Front caliper assembly
20 Caliper
21 Front bracket
22 Bolt - 4 off
23 Screw
24 Bolt - 4 off

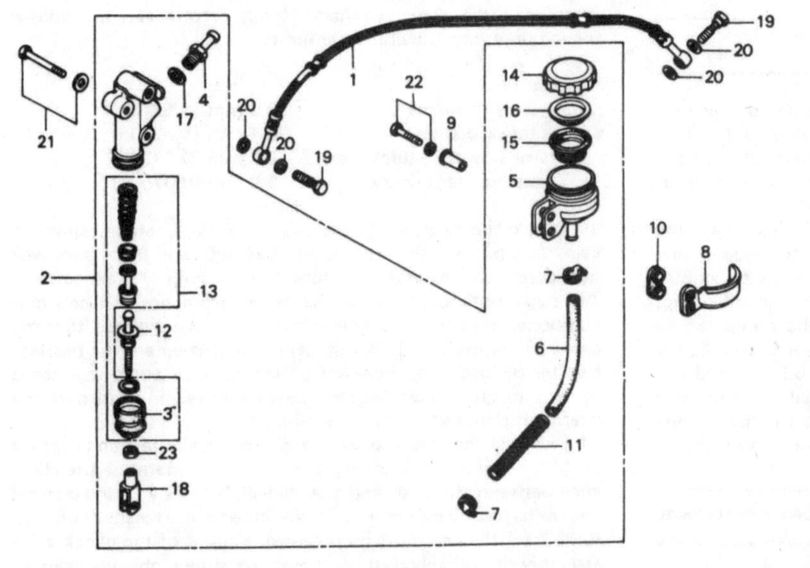

Fig. 8.4. Rear brake master cylinder (F2)

1 *Rear brake hose*
2 *Rear brake master cylinder assembly*
3 *Master cylinder rubber boot*
4 *Connector*
5 *Fluid reservoir*
6 *Hose*
7 *Hose clip - 2 off*
8 *Fluid reservoir cover*
9 *Fluid reservoir collar - 2 off*
10 *Rubber master cylinder mount*
11 *Hose guard*
12 *Pushrod*
13 *Cylinder set complete*
14 *Fluid reservoir cap*
15 *Diaphragm*
16 *Diaphragm plate*
17 *Gasket*
18 *Brake rod clevis joint*
19 *Hose bolt - 2 off*
20 *Washer - 4 off*
21 *Bolt and washer - 2 off*
22 *Bolt and washer - 2 off*
23 *Nut*

12 Final drive chain - inspection, lubrication and adjustment

1 With the use of endless, sealed-in lubricant type of chain, as fitted to the F2 and K7, adjustment should become less frequent and lubrication less time consuming. It is, however, still necessary to check for damaged rollers, loose rivets, or missing O-rings. The construction is such that the grease is sealed into the chain and is retained by the O-rings. Check also for kinked, binding, dry or rusty links. If there is evidence of any of these conditions, lubrication is obviously overdue. This must be done only by greasing or by a light coating of engine oil, preferably of a heavy grade. It is not recommended that one of the proprietary aerosol spray can lubricants be used as these may contain solvents which could damage the rubber O-rings.
2 The chain can be adjusted by following the procedure set out in Chapter 5, Section 14. There is now a further aid to chain adjustment and replacement in the form of coloured chain wear labels on the adjuster stirrups. These take the form of a red and a green section. If the red zone aligns with the rear of the swinging arm, after the chain has been adjusted to its correct tension 20 mm/¾ in), then the chain is excessively worn and must be replaced.

13 Final drive chain - renewal

1 With the latest endless chains the process of removal for renewal is considerably more involved and time-consuming. The chain cannot be split unless a chain rivet extractor is used. There is, however, little to be gained from this as further dismantling will have to take place to fit the new chain.
2 Detach the gearchange pedal, the small cover and the left-hand rear crankcase cover. Apply the rear brake to prevent rotation of the final drive sprocket and loosen and remove the sprocket bolt. Slide the sprocket off the shaft and disengage the chain.
3 Remove the chain guard and rear wheel following the procedure in paragraphs 2 to 4 of Section 9, Chapter 4. Remove the bolts securing the bottom of each rear suspension unit to the top of the swinging arm.
4 Unscrew the swinging arm pivot nut and withdraw the swinging arm pivot bolt. The swinging arm can now be pulled clear and the chain disengaged.
5 Install the new chain by reversing the removal procedure.

14 General description - CB750A model

The CB750A model, introduced to the USA only in March 1976, was something of a departure from traditional motor cycling practice. The major break with tradition was in the adoption of an automatic transmission, hence the suffix letter 'A', and variable drive by means of a fluid torque converter. This method of transmission, designated 'Hondamatic', consists basically of two main components; the torque converter and a 2-speed constant mesh gearbox unit.

The torque converter replaces a conventional clutch and provides a source of torque multiplication. It performs the same functions as a normal geared transmission would if it were equipped with numerous different gear ratios which were infinitely variable between two limits. It is driven by a conventional primary drive geartrain, the mainshaft of which is directly connected to the torque converter, via the 2-speed transmission.

The transmission consists of a mainshaft and a countershaft, and a clutch and series of gears on each shaft. The clutches are of the usual hydraulic multi-plate type and are designated 'L' (for low) and 'D' (for drive). Later models use the designation '1' and '2' in place of 'L' and 'D'. The 'L' clutch is fitted to the mainshaft and the 'D' unit on the countershaft. The final drive is by the usual sprocket on the left-hand end of the countershaft.

With a 'different' type of transmission system such as the 750A utilised, it was inevitable that various other components completely unique to the 'A' would have to be evolved. Therefore, whilst the basic engine was very similar to the original CB750, albeit now with wet sump lubrication, there were several innovations to be found only on the 'A' model.

Apart from the transmission, however, the 750A was very much the same machine as the other CB750 models, sharing most of its running components with the then current manual gearbox models. Later models differed slightly in that the original 4 into 1 exhaust system was dropped in favour of a twin silencer arrangement and a new stepped seat was added.

The following Sections deal with those operations which demand a different approach or method when dealing with a specific task or component. Further, where a component is of an unusual design or its purpose and function may not be immediately obvious, then a description and/or diagram will be included. Where specific mention is not made, it may be assumed that any changes made to the original CB750 to create the 750A model, do not materially affect the methods of operation described in the preceding Chapters.

15 Torque converter - method of operation

The torque converter offers torque multiplication by providing varying drive ratios between the driving and driven members. As the driven member speed approaches that of the driving member it ceases to function as a torque converter. It then takes on the role of a fluid coupling device.

The torque converter consists basically of three fans, the turbine, the pump, and the stator, the latter including a one-way clutch. These components are contained in a doughnut-shaped container filled with oil pressurized by the oil pump. The turbine fan is splined to the mainshaft, the pump fan to the input shaft, and the stator fan to the stator shaft. As the engine is run at increasingly higher speed, the oil is forced out of the vanes of the turbine into those of the pump, and so drive is transmitted. The third fan, or stator, is free running in one direction and directs the oil across from the pump back to the turbine.

16 Torque converter - removal, examination and replacement

1 The torque converter, being a basically simple device, should not cause much in the way of problems. Should servicing become necessary, however, the removal procedure is also straightforward.

2 Remove the right-hand front footrest and the rear brake actuating pedal.

3 Loosen and remove the single retaining screw which holds the slotted protector cover. Remove the cover.

4 The case cap will now be visible. The five retaining bolts for this should be removed. Care should be taken that when removing the cap and its gasket, the check valve and its spring, which are situated behind the cap, are not mis-placed.

5 The torque converter case can now be removed. There are six retaining screws to be taken out, to allow the case to be detached.

6 The torque converter should be removed as a complete unit, and then, if further investigation is required, dismantled away from the machine. It may be possible to grasp the torque converter and draw it off the mainshaft/stator shaft by hand but more probably a puller will be required. Honda recommend the use of service tool No. 07934-3930000 which bolts to the torque converter. If this is not available a large universal legged puller may be used. If the latter method is adopted it will be necessary to bolt suitably fabricated straps across two or more sets of opposing torque converter bolts under which the puller legs may be hooked. Whichever method is used care must be taken to protect the end of the protruding mainshaft against which the puller centre screw will bear. A suitable bearing pad can be made from a bolt, the shank of which just fits inside the hollow shaft end.

7 With the torque converter removed, slacken evenly in a diagonal sequence and then remove the ten bolts which pass through the periphery of the turbine (front of the converter) into the torque converter rear cover. Before displacing the turbine note that a line is scribed across the edge of the unit from the turbine to the rear cover. This line is provided to ensure correct re-alignment of the two components on re-assembly. Separate the turbine from the rear cover using a rawhide mallet around the periphery to break the seal between the two components. Withdraw the stator/clutch unit from the centre of the pump and remove the thrust washer and pump from the rear cover.

8 If wear is suspected in the stator assembly it can now be dismantled. The assembly contains the one-way clutch. This consists of a cam and a hub and two side plates. In the notches in the inside surface of the cam there are 8 sets of rollers and springs. To dismantle the stator assembly the circlips, one at each end, must be removed.

9 Wear should be checked for in the stator hub, the side plates and the thrust washers. Honda recommend the following maximum permissible wear limits:

Stator hub:	
Outside diameter	39.9 mm (1.571 in)
Inside diameter	26.1 mm (1.028 in)
Stator side plate thickness	5.9 mm (0.232 in)
Thrust washer thickness	1.9 mm (0.075 in)

10 Check the bearings in the rear cover and torque converter case for pitting or wear. Both bearings are large and well lubricated and should therefore have a long life expectancy. Although both bearings can be driven from position and new components fitted in a similar manner it is suggested that the cases be returned to a Honda Service Agent where the bearings can be pressed out. Renewal of the oil seal which is inboard of the torque converter case bearing is recommended if the bearing is removed or if wear is suspected.

11 Examine the case cap sleeve and the mainshaft and support bush for scoring or other signs of damage. Determine the clearance between the bush and the mainshaft end by taking internal and external measurement. If the clearance exceeds 0.08 mm (0.003 in) the cap must be renewed. Failure of the check valve assembly is unlikely but if the spring shows obvious signs of weakening or distortion it should be renewed.

12 Reassembly of the torque converter is facilitated if the individual components are built up on the projecting mainshaft/stator shaft unit. This eases relocation of the various mating splined components. Lubricate the rear cover bearing and fit the cover to the shaft. If the large diameter cover sealing ring has stretched or broken it should be renewed.

13 When reassembling the stator/one-way clutch assembly the hub must be fitted with the splined side towards the wider pitched side of the stator. The two side plates must be fitted with their grooved sides facing outer-most. During reassembly it is imperative that each component is kept clean of any, even minute, particles of dirt or foreign matter. All the components must be lubricated with clean engine oil during reassembly. When the stator has been reassembled, with the circlips correctly replaced at each end, the assembly should be refitted to the stator shaft. Check the operation of the one-way clutch at this stage. The clutch should only turn in an anti-clockwise direction.

14 The torque converter turbine can now be replaced. Care should be taken to ensure that the line mark on the turbine aligns with that on the torque converter cover. The retaining bolts should all be refitted and torqued into position (1.2–1.6 kg m/8.7–11.6 lb ft). A large drift can be used to assist both the stator assembly and the turbine back into position. The drift, or bearing driver, should be sufficiently large not to damage the shaft and blows must be applied with a soft-faced mallet in a steady and measured manner.

15 The remaining parts; the torque converter case, the check valve and its spring, and the case cap and its gasket, can now be replaced by reversing the dismantling procedure. Finally, the chromed protector plate cover can be refitted.

16 It should be stressed that although the torque converter assembly is a precision component, it is basically a straightforward device to dismantle and reassemble. Therefore, as long as all the components are kept clinically clean and properly lubricated during reassembly there is no reason for the home mechanic not to be able to service the unit satisfactorily. If any doubt is felt about the operation, however, it should be entrusted to your local Honda Service Agent.

17 Transmission and clutches - method of operation

1 The transmission is a 2-speed constant mesh type. The 'D' (or 2) range is for all normal situations that the machine should encounter. The subsidiary range of 'L' (or 1) is for use when pulling away from a standstill and climbing steep hills, or indeed, descending, when a measure of engine braking is

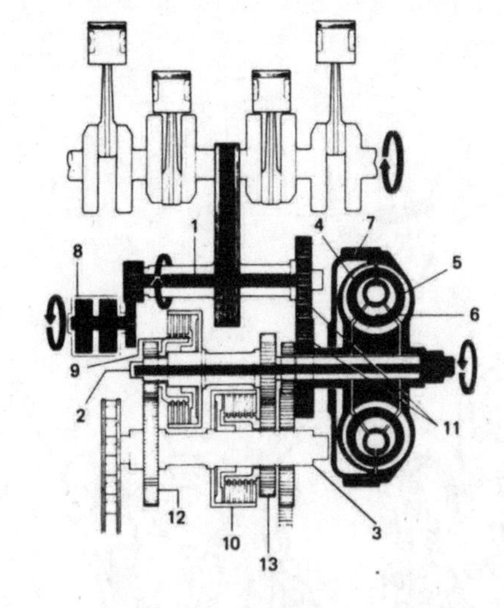

In N range

1. *Primary shaft*
2. *Mainshaft*
3. *Countershaft*
4. *Pump*
5. *Turbine*
6. *Stator*
7. *Torque converter assembly*
8. *Oil pump assembly*
9. *L (or 1) clutch*
10. *D (or 2) clutch*
11. *Primary gears*
12. *Low gear*
13. *Drive gear*

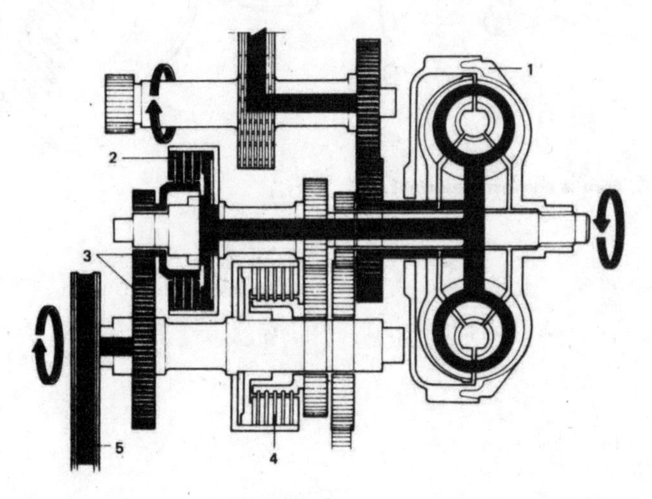

In L (or 1) range

1. *Torque converter*
2. *L (or 1) clutch*
3. *Low gears*
4. *D (or 2) clutch*
5. *Drive chain*

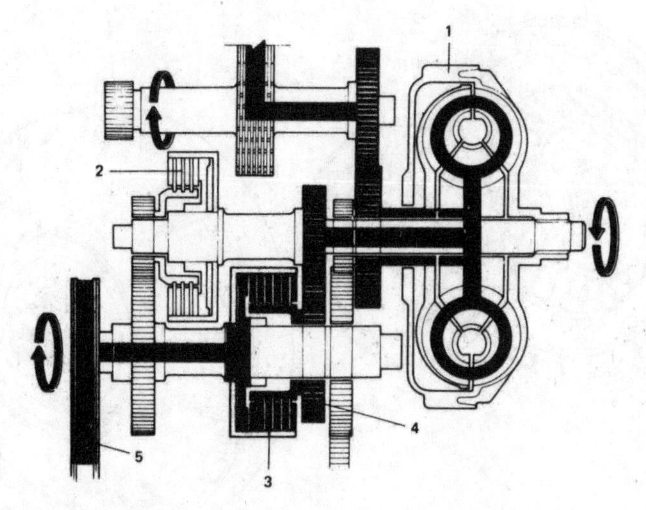

In D (or 2) range

1. *Torque converter*
2. *L (or 1) clutch*
3. *D (or 2) clutch*
4. *Drive gear*
5. *Drive chain*

Fig. 8.5. Automatic transmission - power flow

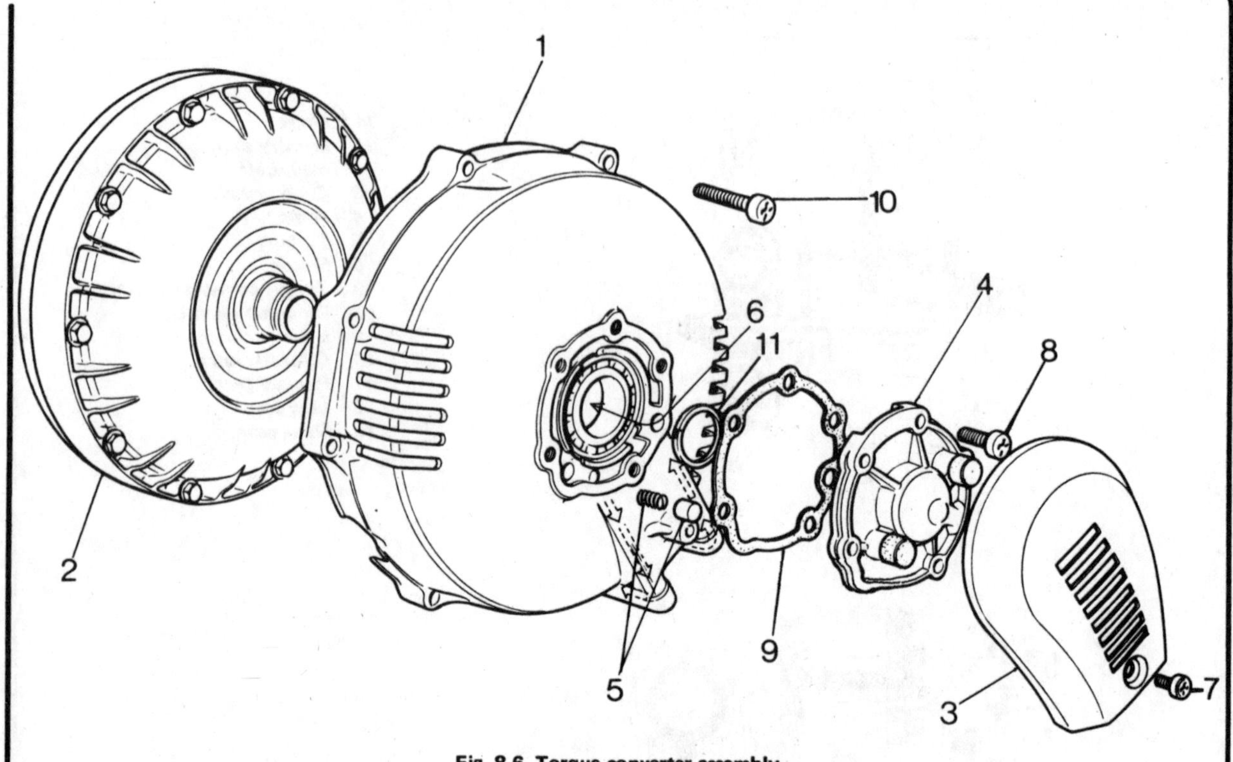

Fig. 8.6. Torque converter assembly

1 Torque converter casing
2 Torque converter
3 Case cap shield
4 Case cap
5 Check valve assembly
6 Bearing
7 Screw
8 Screw - 5 off
9 Gasket
10 Screw - 6 off
11 Bush

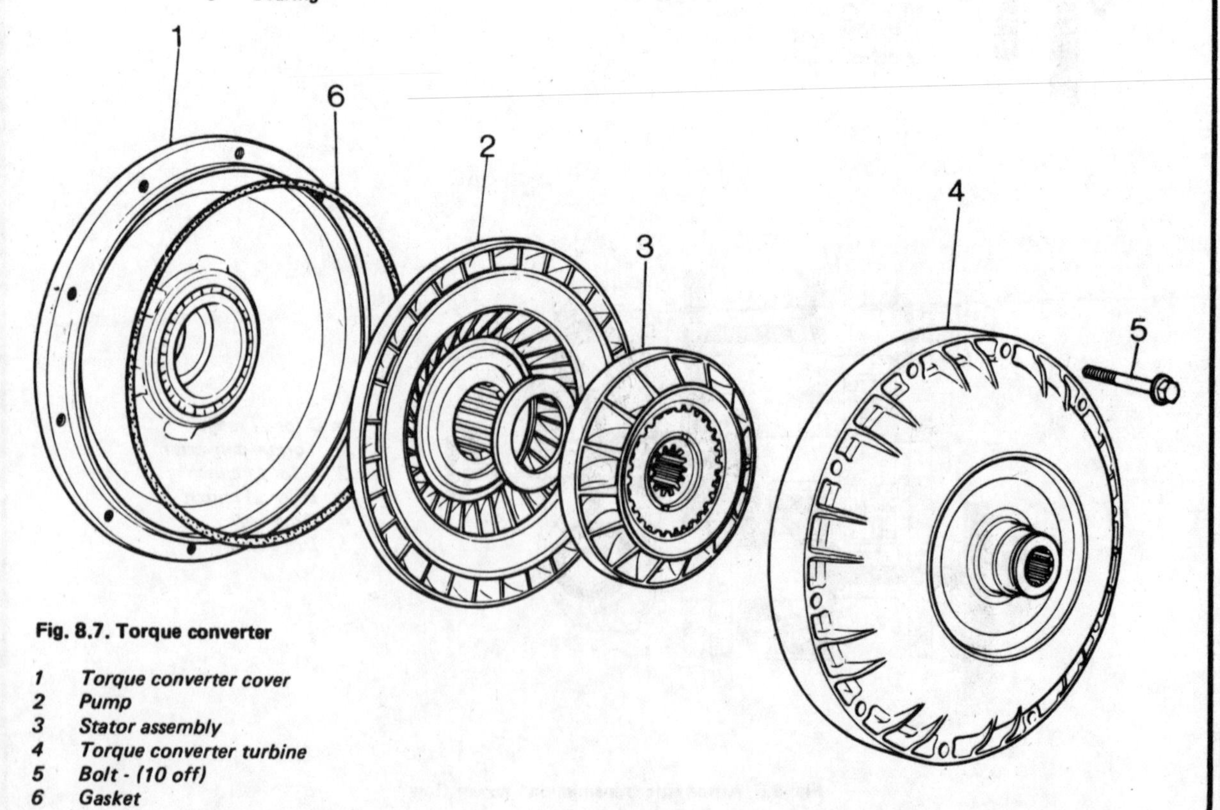

Fig. 8.7. Torque converter

1 Torque converter cover
2 Pump
3 Stator assembly
4 Torque converter turbine
5 Bolt - (10 off)
6 Gasket

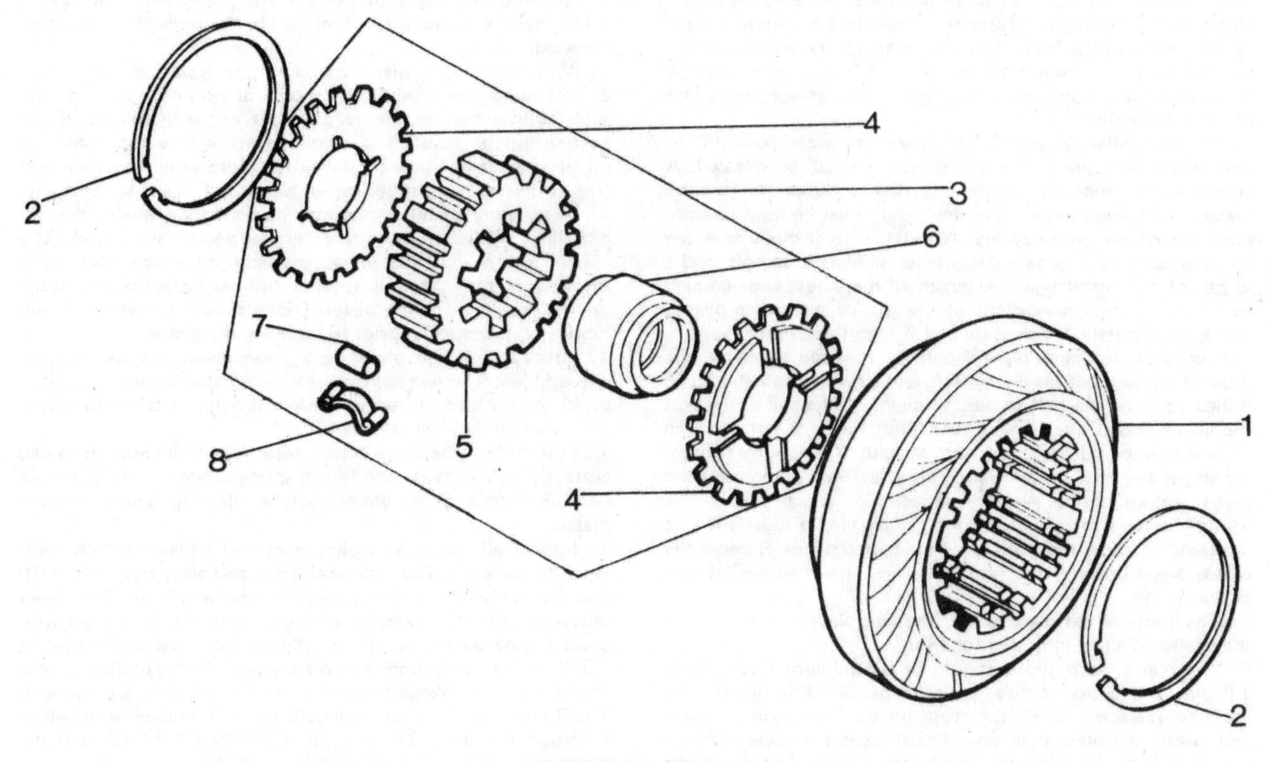

Fig. 8.8. Stator assembly

1 *Stator*
2 *Circlip - 2 off*
3 *One-way clutch assembly*
4 *Side plate - 2 off*
5 *Cam*
6 *Hub*
7 *Roller - 8 off*
8 *Spring - 8 off*

most beneficial. Further to these ranges, a third position is available; that of 'N' for neutral. Control of this transmission is achieved by two built-in hydraulic clutches and the normal gearchange pedal.

2 As the engine is started the oil pump supplies oil pressure to the torque converter. With the transmission in the 'N' range, no oil is sent to either clutch. There is, therefore, no power transmitted from the mainshaft to the countershaft and hence no movement of the machine. When the 'L' (or 1) range has been selected, the relevant clutch, the one fitted to the mainshaft, is engaged. When operating in the 'D' (or 2) range the second, countershaft mounted, clutch is engaged.

3 The flow of power from the engine, therefore, follows a specific route. It flows from the torque converter to the mainshaft through whichever clutch is engaged, on to either the low or drive gears and from there to the countershaft which carries the final drive sprocket.

4 The clutches themselves are of the normal hydraulic multi-plate type; the 'L' unit containing four friction and four plain plates and the 'D' clutch six of each type of plate.

18 Transmission and clutches - removal, examination and replacement

1 To enable transmission/clutch examination and overhaul to be carried out, it is necessary to dismantle the major part of the engine. The stripdown sequence, as described in Chapter 1 can be followed. The engine parts have, of necessity, to be removed, once the engine has been extracted from the frame, as the crankcase halves have to be separated. The complete top-end assembly, comprising the cylinder head, barrel block, pistons and cam chain tensioner must be dismantled. Also, the torque converter case, oil pump assembly, gearchange and automatic neutral return system assemblies, and regulator valve mechanism must be removed. Torque converter removal is discussed in Section 16 as is removal of the oil pump/gearchange assembly in Section 21.

2 With the engine design modified, to accept automatic transmission and wet sump lubrication, the crankcase securing bolt arrangements have changed. Seven upper crankcase securing bolts and twenty-three lower half securing bolts must be removed.

3 With the crankcases separated the mainshaft can be raised and the primary sprocket and mainshaft gear assembly can be removed. Remove the stator and input shafts as complete assemblies. The countershaft and its gears can also now be removed as a complete assembly.

4 Remove the mainshaft bearing, thrust washer, 'low' gear and its needle bearing and second thrust washer, to allow the 'low' clutch to slide off the mainshaft, ensuring its locating dowel has been displaced. The 'drive' clutch can be removed from the countershaft in a similar manner, again noting the positioning of bearing, thrust washers and oil seals and displacing the locating dowel.

5 Examine the two clutches and their associated geartrain components now they are separated from the mainshaft and countershaft. The gear teeth should be inspected for any signs of excessive wear, cracks or scoring. The shaft bearing surfaces should be checked for scoring, scratches or excessive wear, and the bearings themselves examined for damaged rollers and roughness on rotation. If any of these faults are apparent replace the offending item(s) before commencing rebuilding.

6 Each of the clutches should also be examined and, if necessary, new components fitted. Measure the clearance between the clutch end plain plate and the top friction plate by inserting a feeler gauge blade. The service limit is 0.5–0.8 mm (0.020–

0.031 in). If the clearance is greater than this amount a new clutch end plate should be selected (from the table under Clutch Specifications at the front of this Chapter) so that when installed the correct clearance is restored. Note that when inserting the feeler gauge blade, care must be taken not to damage the surface of the friction disc.

7 To dismantle the clutch, to check for wear generally or replace the end plate, the clutch piston must be released. A Honda special tool, No. 07960-6120000, a clutch spring compressor, or a home-made equivalent tool, must be used to compress the clutch return spring. A suitable home-made tool can be constructed by using a six-inch (or similar) long bolt, and a length of flat metal plate. A length of metal just wide enough to fit the internal dimensions of the top of the clutch piston, and approximately 1 inch wide and ¼ inch thick should suffice. The ends of the metal plate should be rounded to match the internal contours of the clutch unit, and a central hole should be drilled to allow the bolt to pass through. To the end of the bolt should be added one or two nuts. With the tool installed, the plate should be pushed down onto the clutch piston, by tightening down the nuts on the bolt. This effectively compresses the clutch return spring allowing the circlip, which retains the spring, to be removed. A second, larger circlip must then be displaced to allow the removal of the clutch plates. Remove the clutch plates one at a time noting their relative positions for ease of reassembly.

8 The friction and plain plates, the end plate, and the return spring should all be inspected for wear.

9 The plates, both types, should be a minimum thickness of 1.9 mm (0.075 in). If they are worn beyond this figure, they must be replaced. Check the teeth on the end plate for wear and check the plate thickness. The thickness is dependent on the plate type, identified by a stamped number (see paragraph 6, this Section). The clutch return spring should have a free (uncompressed) length of no less than 36.0 mm (1.42 in). If it has taken a permanent set and is less than this, it too must be renewed.

10 With the clutch piston removed the inside of the clutch drum can be examined. There should be no undue wear on the shaft splines nor on the clutch plate engaging splines. Slight burring can be rectified by careful work with a small file. The oil passages in the base of the clutch drum should be clear and free from any restrictions which could lead to clogging.

11 Reassembly of the clutch may be done by reversing the dismantling procedure, ensuring certain points are noted. The clutch piston must be seated properly; to ensure this apply pressure as the piston is rotated anti-clockwise in the clutch drum. The large clutch piston O-ring should be replaced, and lubricated thoroughly, prior to installing the piston.

12 Lubricate all the plates with clean engine oil as they are reassembled. The two split oil seal rings fitted behind the clutch drum should also be renewed. Care should be taken to ensure their gaps are 180° apart when fitted.

13 Ensure that the large clutch plate retaining circlip is seated correctly and likewise the return spring retaining circlip should be fully home in its groove before releasing tension on the plates.

14 Inspect all the transmission gears and replace if excessively worn or damaged. The various clutches and gears and their shafts can be reassembled following the reverse of the stripdown sequence. All the transmission parts must be lubricated with clean engine oil during the rebuilding stage, and a thin coat of liquid gasket compound should be applied to the mating surface of the lower crankcase half when the two halves are rejoined. The 8 mm crankcase securing bolts should be tightened down to a torque figure of 2.0–2.5 kgf m (14.5–18 lbf ft), and the 6 mm bolts to 1.0–1.4 kgf m (7.2–10 lbf ft).

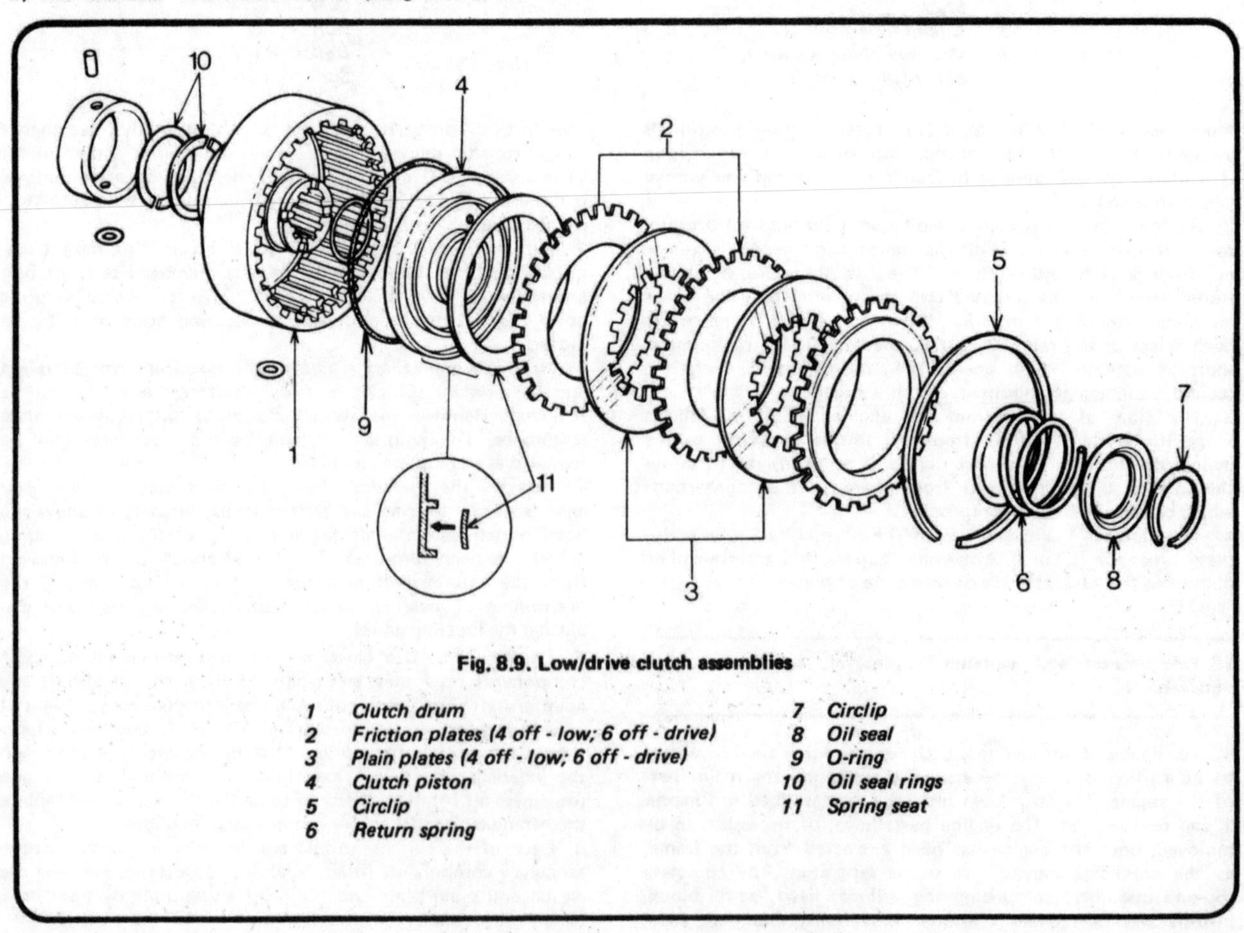

Fig. 8.9. Low/drive clutch assemblies

1 *Clutch drum*
2 *Friction plates (4 off - low; 6 off - drive)*
3 *Plain plates (4 off - low; 6 off - drive)*
4 *Clutch piston*
5 *Circlip*
6 *Return spring*
7 *Circlip*
8 *Oil seal*
9 *O-ring*
10 *Oil seal rings*
11 *Spring seat*

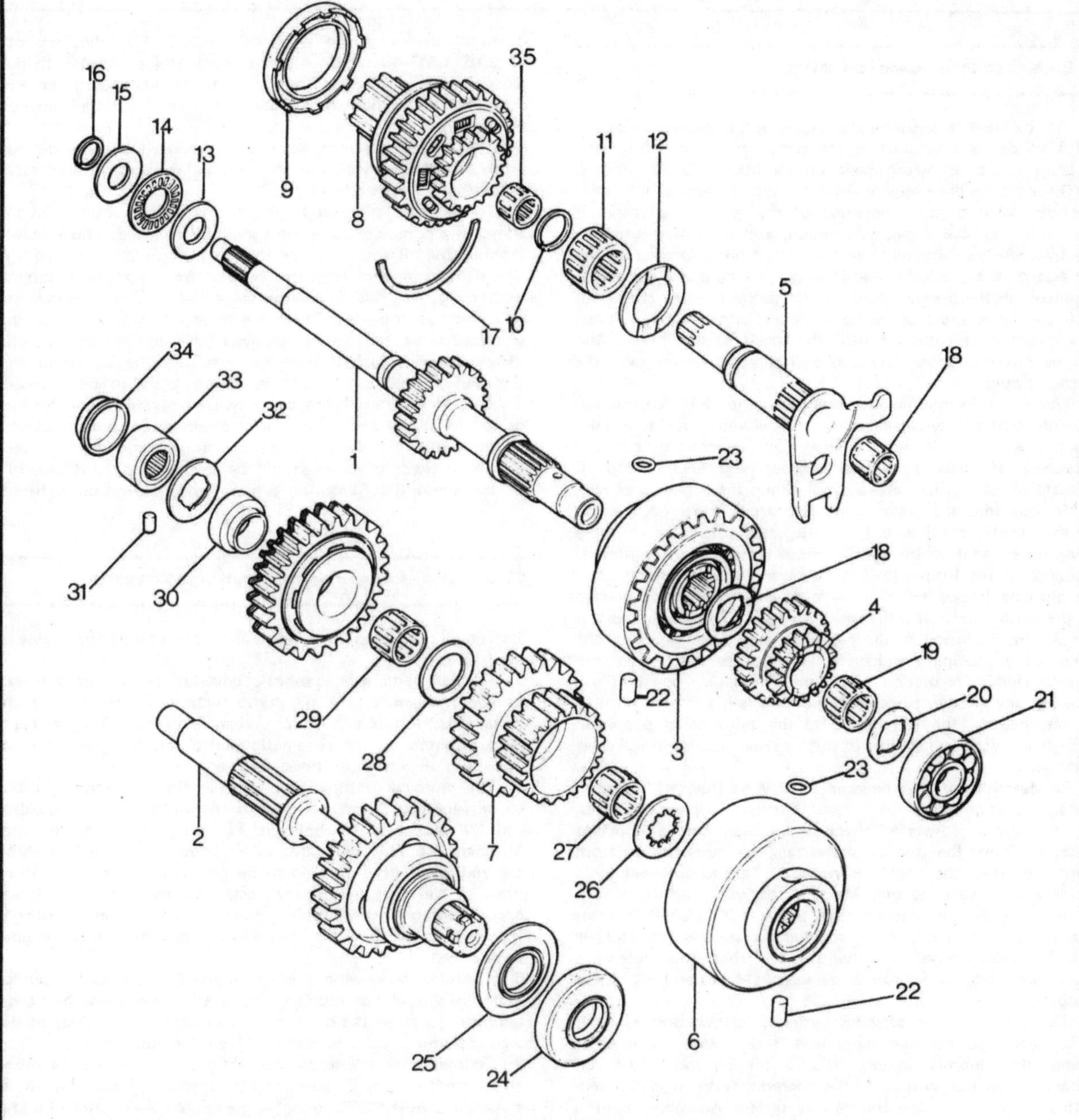

Fig. 8.10. Gearbox components

1	*Mainshaft*	19	*Needle roller bearing*
2	*Countershaft*	20	*Splined washer*
3	*Low clutch*	21	*Bearing*
4	*Low gear*	22	*Dowel pin - 2 off*
5	*Stator shaft*	23	*Retaining ring - 2 off*
6	*Drive clutch*	24	*Oil seal*
7	*Drive gear*	25	*Belville washer*
8	*Input shaft*	26	*Splined washer*
9	*Bearing seal*	27	*Needle roller bearing*
10	*Circlip*	28	*Washer*
11	*Needle roller bearing*	29	*Needle roller bearing*
12	*Thrust washer*	30	*Shouldered bush*
13	*Thrust washer*	31	*Dowel pin*
14	*Needle roller washer*	32	*Splined washer*
15	*Thrust washer*	33	*Spacer*
16	*Circlip*	34	*End cap*
17	*Bearing location ring*	35	*Needle roller bearing*
18	*Needle roller bearing*		

19 Engine and transmission lubrication

1 The oil used to lubricate the engine and transmission of the 750A model is contained in the sump, as distinct from the separate oil tank arrangement on all other CB750 models.
2 Oil is fed to the engine and Hydromatic system by a tandem trochoid rotor pump. The inner of the two pump rotors is responsible for the engine lubrication, and the outer furnishes the transmission with oil. The complete unit is located on the left-side of the crankcase and is driven by the pump drive gear mounted on the primary drive shaft. The inner rotor is integral with the drive shaft operating in series with the outer rotor. The outer rotor rotates at 4/5 the speed of the inner rotor, forcing out oil, under pressure, each time it passes over the discharge port.
3 There is a further feature, unique to the 750A, of the lubrication system; the regulator valve. This valve works, in conjunction with the oil pump, to maintain a constant oil pressure regardless of changes in engine speed and engine load. It consists of two springs and a spool. When the engine is at rest, in 'N' position, the stator shaft is bearing down on the top of the regulator valve, but not depressing it, due to there being no oil sent to the torque converter. As soon as there is a change in the torque loading however, as would occur when the throttle is opened, there is a downward pressure exerted by the stator shaft onto the regulator valve. This would, were it not for the inclusion of the regulator valve springing, have the effect of producing a sudden surge of oil to the torque converter. Under regulation the flow maintains constant by allowing any sudden surges to be controlled, and excess oil to be returned to the sump, due to the valve being depressed enough to permit oil to flow through a relief passage in the valve body side.
4 Besides the usual oil pressure relief valve inserted in the oil passage plate, there are two further, separate valves.
5 The torque converter check valve maintains a constant pressure from the torque converter and prevents it from emptying when the engine is running. This is achieved by a spring loaded valve closing whenever the oil pressure is low. As soon as the oil pressure rises above 1.0 kg/cm^2 (14 psi) the spring force is overcome, and the excess oil is free to flow out. For information on dismantling and renovation reference should be made to Section 2, paragraphs 4, 10 and 13 of this Chapter.
6 The manual valve provides hydraulic clutch operation. It is connected to the gearchange lever and enables the rider to change the transmission into 'N', 'D' (or 2), and 'L' (or 1), depending on the position of the operating lever. It is situated, with the automatic neutral return system assembly, behind the left crankcase cover.

20 Oil filter and screen - cleaning and renewal

1 Drain the oil from the engine by removing the drain bolt situated on the left underside of the sump.
2 To extract the filter element, the filter housing must be removed. This is done by withdrawing the centre bolt, taking care to place a receptacle below the filter unit to catch the escaping oil.
3 The old filter should be discarded and a new one fitted. The O-rings on the oil filter bolt and the oil filter housing should also be renewed. Do not overtighten the centre bolt, the correct torque setting is 2.7–3.3 kgf m (19.5–24 lbf ft), as thread damage can easily occur. The oil filter housing and crankcase drain plug should be refitted.
4 Refill the crankcase with 4.0 litres (8.4/7.0 US/Imp pt) of SAE 10W–40 motor oil and check the oil level with the dipstick now provided. The oil level should be between the upper and lower limit marks with the dipstick inserted but **not** screwed in.
5 After the engine has been started and allowed to idle for a few minutes recheck the oil level, and replenish if necessary, and check for any oil leaks.
6 To facilitate removal of the strainer screen situated in the sump, the complete sump unit must be removed. This involves draining the oil and removing the fourteen sump retaining bolts. The sump unit can then be lowered from position complete with its gasket. The regulator valve and oil distribution plate will then be apparent. There are three retaining bolts for the regulator valve; with these removed the valve assembly can be lifted clear. A further thirteen bolts must be slackened and removed to allow the oil distributor plate to be released.
7 The oil strainer screen can now be separated from the base of the oil pump and washed in a clean solvent such as petrol. If the screen is damaged or broken it must be replaced.
8 The procedure above should be reversed for the reassembly of the sump unit, ensuring a new sump case gasket is fitted.

21 Oil pump - dismantling, examination and reassembly

1 The oil pump can be removed and replaced with the engine in the frame.
2 The left-hand side crankcase cover must first be removed to allow access to the oil pump retaining screws. There are three crosshead screws which, when removed, allow the complete assembly of the automatic neutral return system, manual valve, and oil pump unit to be removed.
3 The removal of three bolts enables the oil pump assembly to be separated from the rest of the left-hand case assembly and the gasket to be removed. Pull out the oil passage pipe.
4 There are now two bolts which must be released to allow the end cover of the pump to be removed. Lift out the outer rotor of the first pump, then take out the dowel pin which passes through the oil pump shaft, so that the inner rotor is free to slide off the shaft. The body of the first pump can now be removed.
5 A similar arrangement is employed for the second pump. The pin should be displaced, once the inner rotor has been removed, to allow the outer rotor to slide free. The body of the second pump can also be removed from the pump shaft.
6 Following the measurements laid down in the Specifications at the front of this Chapter, and the method of measurement as stated in Chapter 2, Section 14, paragraphs 8–11, the oil pump can be checked, and if necessary, renovated.
7 Reassemble the pump(s) by reversing the dismantling procedure. Make sure that all the components are well lubricated, especially the area between the inner and outer rotors of each pump.

22 Regulator valve - removal

1 The regulator valve, as previously stated in Section 5, paragraph 3 of this Chapter, provides a method of maintaining constant oil pressure regardless of changes in engine speed and engine load.
2 With the engine oil drained and the sump pan removed, the regulator valve and oil passage distribution plate are accessible. Three bolts retain the regulator valve. It is recommended by Honda that the valve unit is not dismantled, and as such no service information is given. If the unit is thought to be suspect, therefore, the only choice of action is to seek the advice of your local Honda Service Agent.

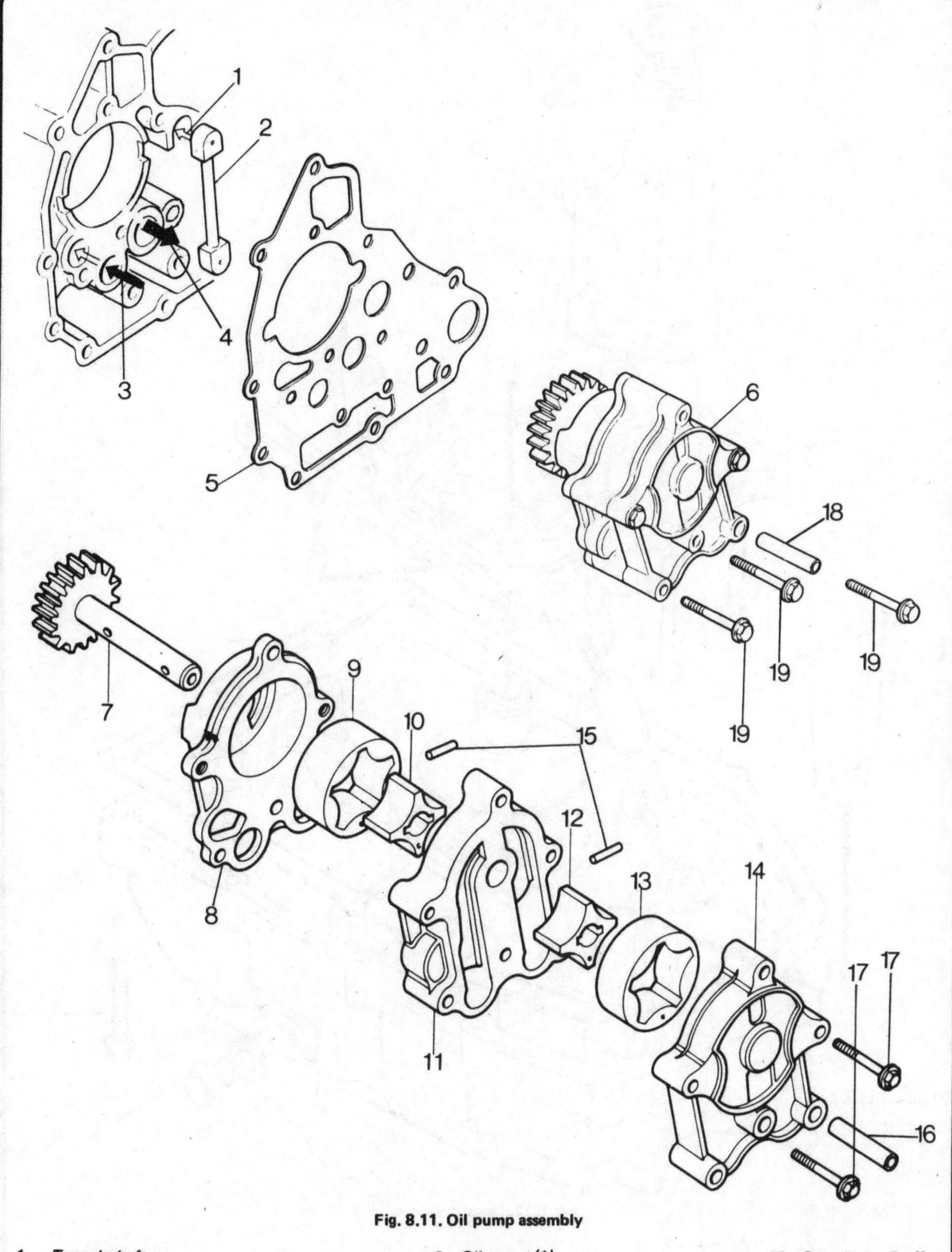

Fig. 8.11. Oil pump assembly

1. *To mainshaft*
2. *Oil passage pipe*
3. *To engine and transmission*
4. *To oil pump*
5. *Left-hand casing gasket*
6. *Complete oil pump assembly*
7. *Oil pump shaft*
8. *Oil pump (1)*
9. *Outer rotor*
10. *Inner rotor*
11. *Oil pump body (2)*
12. *Inner rotor*
13. *Outer rotor*
14. *Oil pump cover*
15. *Dowel pin - 2 off*
16. *Spacer*
17. *Bolt - 2 off*
18. *Spacer*
19. *Bolt - 3 off*

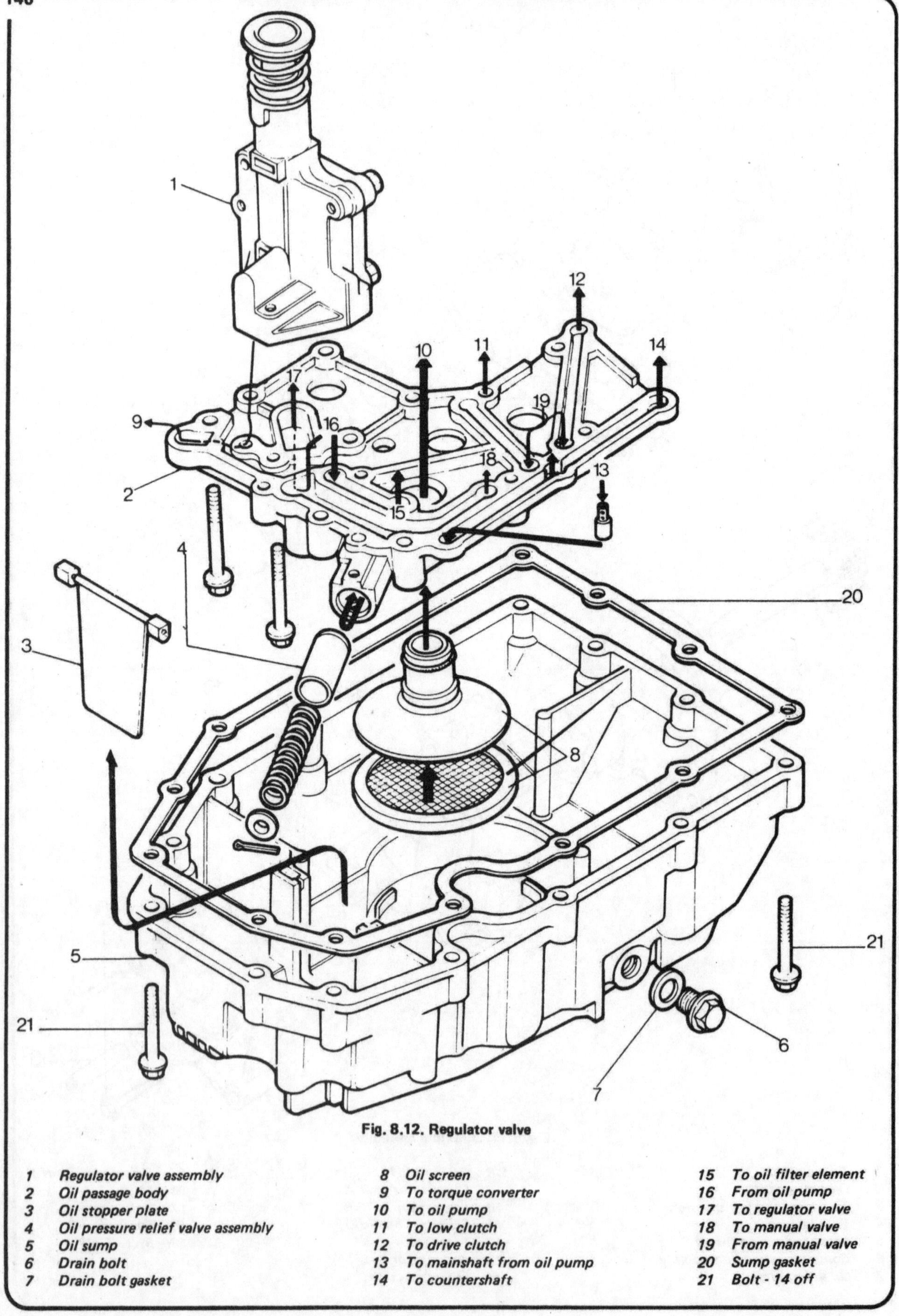

Fig. 8.12. Regulator valve

1	*Regulator valve assembly*	*8*	*Oil screen*	*15*	*To oil filter element*
2	*Oil passage body*	*9*	*To torque converter*	*16*	*From oil pump*
3	*Oil stopper plate*	*10*	*To oil pump*	*17*	*To regulator valve*
4	*Oil pressure relief valve assembly*	*11*	*To low clutch*	*18*	*To manual valve*
5	*Oil sump*	*12*	*To drive clutch*	*19*	*From manual valve*
6	*Drain bolt*	*13*	*To mainshaft from oil pump*	*20*	*Sump gasket*
7	*Drain bolt gasket*	*14*	*To countershaft*	*21*	*Bolt - 14 off*

23 Manual valve/automatic neutral return system - dismantling, examination and reassembly

1 Refer to Section 5, paragraph 6, of this Chapter for a description of the manual valve.

2 The automatic neutral return system is a method of ensuring unintentional starts in either gear position are not made. The operation of the system is dependent upon the actuation of the sidestand. With the sidestand deployed, the gearchange will remain in neutral. This is achieved by the neutral arm engaging with a pin on the manual valve plunger, and preventing the manual valve from fulfilling its normal role of changing the gear position. As soon as the sidestand is raised, the neutral arm is lowered, allowing gear-changing to recommence.

3 With the neutral arm in a horizontal mode, the manual valve plunger is at its highest position. This is the 'N' position. When the neutral arm and manual valve plunger are partially lowered, the 'L' (or 1) range is engaged. With the neutral arm fully lowered, and the valve plunger at the bottom of its stroke, the 'D' (or 2) range is operative.

4 Remove the gearchange pedal, the neutral return arm, and the crankcase side cover. The gearchange inner actuator and outer housing are removed as a unit. Unhook the neutral keeping spring and remove the spindle to which it is attached. As this is being removed displace the heavier spring on the inside end of the spindle. The neutral selecting arm should then be drawn off the gearchange pivot shaft, and the shaft itself then removed. Slacken and remove the nut retaining the ratchet guide, gearchange stopper arm and spring. Remove the gear position indicator switch, which is retained by one crosshead screw.

5 The manual valve can now be removed by unscrewing its four retaining bolts. Note the valve assembly gasket, and exercise care not to damage it. Examine the valve body and plunger shaft for signs of wear.

6 All the components should be visually checked for signs of wear, especially the springs, and replaced if unduly worn. To reassemble the components, follow the reverse of the procedure given above.

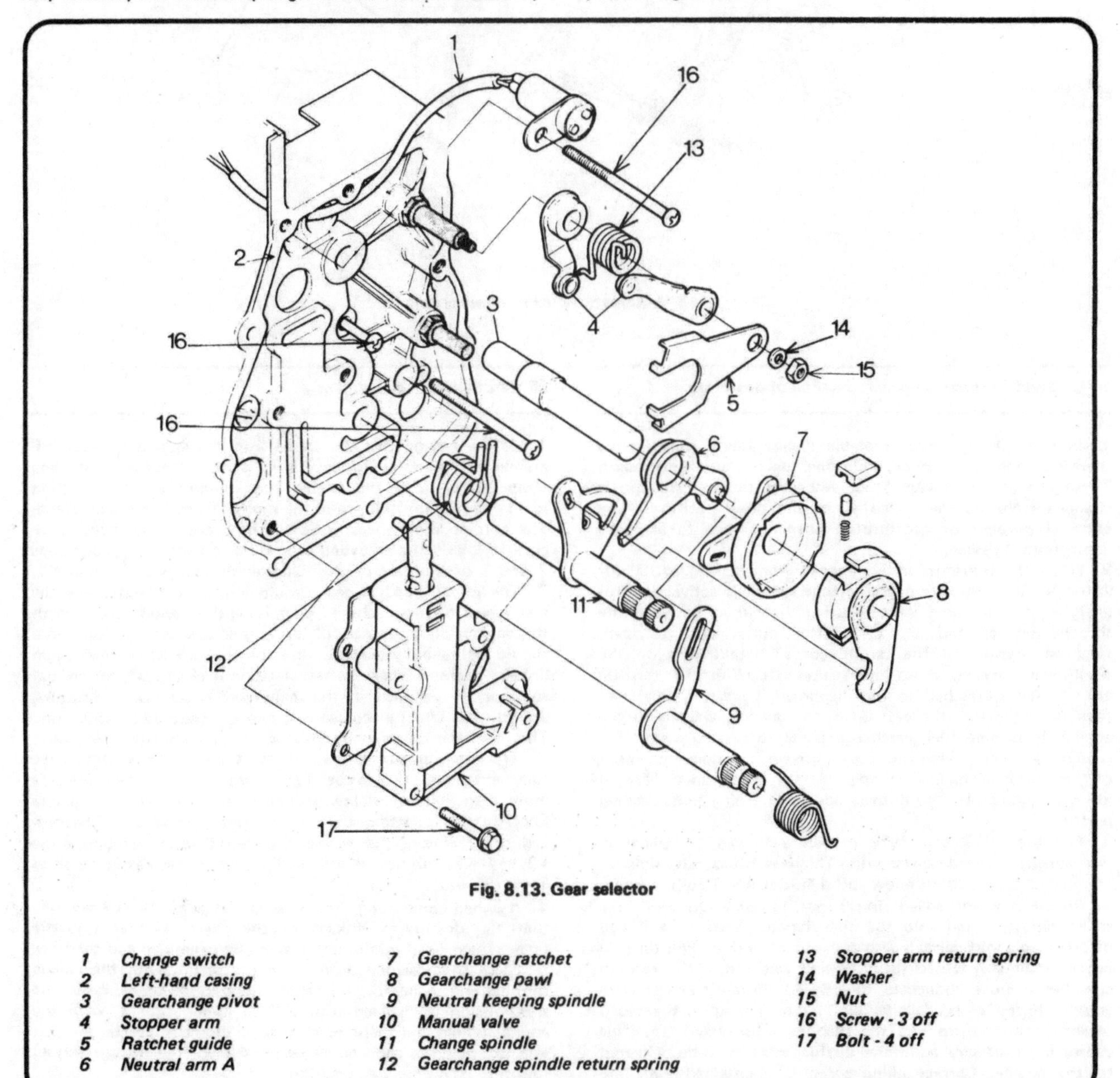

Fig. 8.13. Gear selector

1 Change switch
2 Left-hand casing
3 Gearchange pivot
4 Stopper arm
5 Ratchet guide
6 Neutral arm A
7 Gearchange ratchet
8 Gearchange pawl
9 Neutral keeping spindle
10 Manual valve
11 Change spindle
12 Gearchange spindle return spring
13 Stopper arm return spring
14 Washer
15 Nut
16 Screw - 3 off
17 Bolt - 4 off

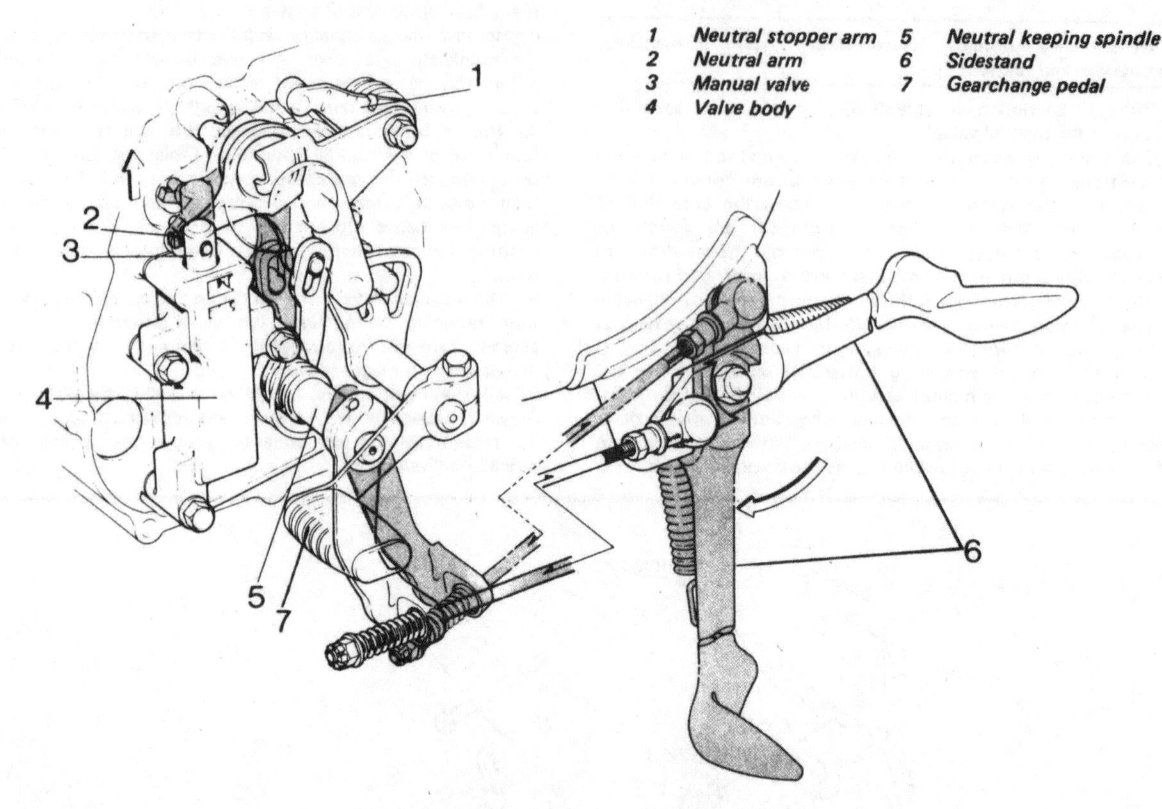

Fig. 8.14. Automatic neutral return system

24 Carburettor throttle opener - method of operation

1 In order to maintain a stable engine idling speed when there are changes in torque, additional fuel mixture is necessary. These changes in torque occur whenever the transmission is changed from neutral to either of the drive positions. The essential purpose of the throttle opener is then to act as a compensating device.

2 When the transmission is changed from neutral to 'D' (or 2) or 'L' (or 1) position, the change switch is activated. This energises the solenoid valve, causing the circuit between the throttle opener and the carburettor outlet side to open. Negative pressure at the carburettor air intake acts on the diaphragm, causing it to move upwards. With the throttle *opening* lever attached to the diaphragm by a rod, this lever then pulls the throttle lever up; supplying fuel and air to the engine. Returning the gearchange pedal to neutral causes the switch to cancel. When in the 'N' range, the upper chamber of the throttle opener is open to the air cleaner. Normal idling is regained by the throttle opener returning to its original position.

3 For the 1977 and 1978 models a further complication was added; a speed sensor unit. This was fitted ostensibly to counteract problems of a low speed stutter when pulling away.

4 As the machine speed rises above 11 mph (20 kph), the speed sensor, fitted into the speedometer head, is switched off. The solenoid valve is also put out of action. The throttle opener lever is returned to its original position as the vacuum chamber is now connected directly to the air cleaner. Conversely, if the speed falls *below* 11 mph (20 kph), the speed sensor is turned on and the diaphragm is pulled up. This causes the necessary additional air/fuel mixture to be delivered to the engine. Correct idling speed is maintained with the throttle lever returning to its original position.

25 Carburettors - adjustment

1 Before setting the idle speed, run the engine for approximately ten minutes to ensure it is at its normal operating temperature. With the engine running and the transmission in the neutral position, raise the machine onto its centre stand. The correct idling speed is 950 rpm. If the idling speed is incorrect follow the adjusting procedure in Section 7, paragraphs 2 and 3 of the part of this Chapter dealing with F2 and K7.

2 The idling speed in gear should also be checked. Lock the rear wheel of the machine by applying the parking brake. With the engine idling, engage 'L' (or 1) gear position. A visual note should be taken to ensure that the stall preventer diaphragm linkage operates when the transmission is changed from neutral to 1 or 'L' position. If the diaphragm is operating correctly, the 'in gear' idling speed can be checked. It should be 1000 rpm. The throttle stop screw must be used if adjustment is necessary.

3 The fast idle speed, that is with the choke operating knob fully extended, should be 1750 rpm. If the engine idles at more than 750 rpm below or above 1750 rpm, with the choke knob in this position, then adjustment is necessary. The procedure in Section 7, paragraph 5 of this Chapter, relating to the F2 and K7, should be followed to enable this adjustment to be carried out.

4 If, when conducting the idle speed 'in gear' test, it was noted that the diaphragm linkage of the stall preventer (throttle opener) was not functioning, it must be examined and checked.

5 Raise the machine onto its centre stand, with the engine running and in neutral, and apply the parking brake. When the transmission is changed from 'N' to either 1 or 2, with the engine idling, the diaphragm linkage should move to the fast idle position. As soon as 'N' is re-selected, the linkage should return to its normal idle position.

6 On 1977 and 1978 models, which are equipped with the

speed sensor unit, a simple check can be carried out to ascertain the exact functioning of the throttle opener diaphragm linkage. Remove the solenoid valve from the circuit and connect a test light in its place. The test light should be placed in such a position as to be visible when the machine is ridden, for example taped between the two instrument heads. Ride the machine and note when the light is extinguished and when it is re-lit. The light should go out at speeds over 14.5 mph (23 km/h) and come back on at speeds below 10.5 mph (17 km/h). This shows the diaphragm is acting correctly. Disconnect the test light and re-install the solenoid valve.

7 If, during the visual examination (the only test possible on 1976 models) or the test light check, the stall preventer (throttle opener) is found to be inoperative then the vacuum in the system must be checked. Check all the vacuum lines for leakage and correct routing, a pinched or trapped air line can be an effective barrier to the flow of air in the system. If the fault cannot be traced this way then there may be a loss of vacuum elsewhere. Lock the back wheel with the parking brake, and with the engine running in gear, disconnect the vacuum tube at the throttle opener. Check for vacuum at the diaphragm with a vacuum gauge. The correct figure should be 13.8 in Hg (350 mm Hg) min.

8 If there is no vacuum then the voltage across the solenoid valve should be checked. If there is no voltage, then suspect either a faulty earth circuit, which would affect the speed sensor (not 1976 models) and/or the change switch, or a faulty power circuit. If these checks are passed successfully, and a new solenoid valve is installed, then inspect the check valve and recheck the vacuum line to the carburettor, looking for any obstruction.

9 The operation of the solenoid valve can be checked by a simple test. With the solenoid operating correctly there should be air flow from the vacuum line from the throttle opener to the air cleaner. Conversely, if the vacuum line to the carburettor is disconnected, and the tube inserted into the tester's mouth, there should be no air flow from the air cleaner when a sucking force is applied. Similarly, no air should flow from the throttle opener to the tester's mouth, but it should flow from the mouth to the air cleaner when the solenoid is de-energised.

10 To test the check valve in the system simply check the flow of air through the vacuum tubes each side of the check valve. The tube connecting the solenoid to the check valve should be disconnected, and inserted into the tester's mouth. Air should flow from the mouth to the tube connecting the valve to the carburettor. It should not, however, flow from the carburettor to the solenoid.

26 Parking brake - method of operation

1 With any automatic transmission, 'creep' is an inevitable nuisance. Due to creep, the rear wheel will always try to turn with the machine stopped, but with the engine still running. The parking brake is used to prevent this creeping.

2 The brake is cable operated from a push/pull knob situated below the left side of the petrol tank, in conjunction with the normal foot pedal.

3 The brake pedal is fitted to the usual splined shaft with a ratchet and pawl system to engage and disengage the brake.

4 When the push/pull knob is pulled out, the ratchet lever rotates over the ratchet pawls. With the brake not in operation, the pawls are held in place in their respective notches by the ratchet lever. As soon as the brake pedal is depressed, the lever is rotated, and the pedal is held down by the engagement of the ratchet and pawls.

5 When releasing the brake, the ratchet lever is disengaged by means of a return spring. The instrument console warning light is extinguished by the ratchet lever breaking the parking switch contact.

6 To put the parking brake into operation requires the rider to pull out the under-tank knob, and depress the rear brake pedal. A visual check that the brake is on, is the illumination of the 'Parking' warning light on the right-hand instrument dial.

7 To release the brake, the knob should be returned to its former position by depressing the centre button, the brake lever pushed down again, ensuring it then returns to its normal mode, and a check made that the 'Parking' light has been extinguished.

27 Parking brake - examination and adjustment

1 To examine the ratchet mechanism of the parking brake, the brake pedal must be removed and the cover, retained by two crosshead screws, displaced.

2 Make a visual inspection of the condition of the ratchet lever and mechanism and the ratchet pawls. The parking brake should be locked on each time the pedal moves approximately ¾ inch. If this is not the case, suspect damage or wear in the mechanism and renovate accordingly.

3 Lubricate all mating surfaces in the assembly, and make sure the ratchet pushes up to make contact with the 'Parking' light switch.

4 The 'Parking' brake cable should have a free play of 2.0 mm (0.8 in). If adjustment is necessary, loosen the locknut and turn the adjuster nut as required to restore correct free play. The adjustment nuts are situated at the top of the ratchet assembly cover. The parking light switch is fitted alongside. Adjustment of the latter is correct when the switch is fully screwed in.

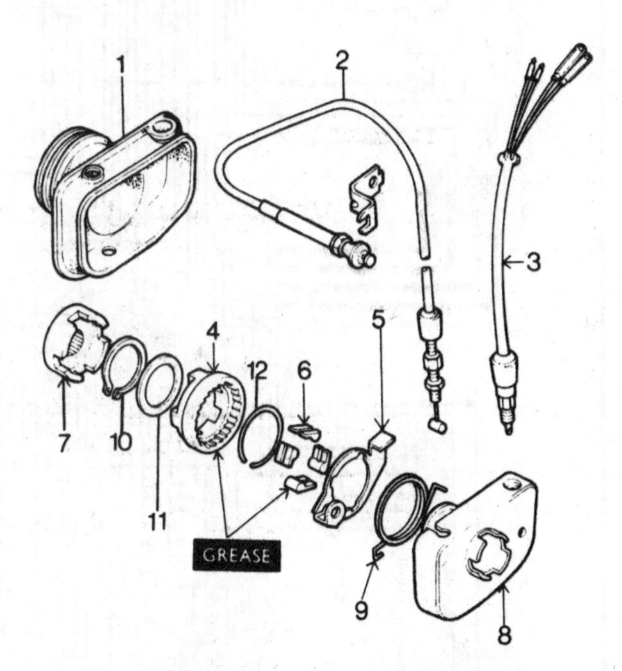

Fig. 8.15. Parking brake system assembly - '76 model

1	*Ratchet cover*	7	*Rear brake joint*
2	*Parking brake cable*	8	*Ratchet base*
3	*Parking switch*	9	*Return spring*
4	*Ratchet case*	10	*Circlip*
5	*Ratchet lever*	11	*Seal*
6	*Ratchet*	12	*Circlip*

**The parking brake assembly fitted to later models is similar.*

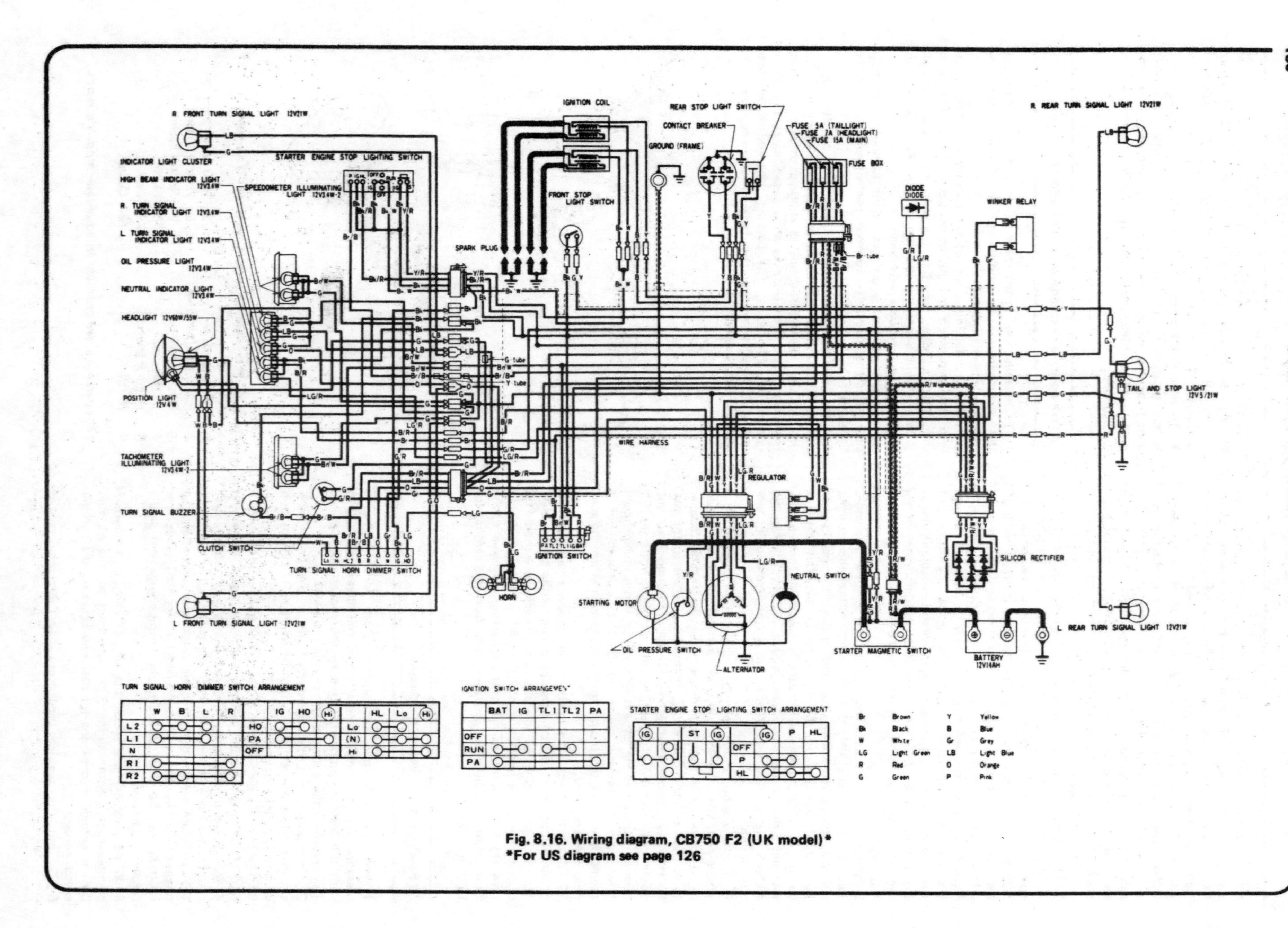

Fig. 8.16. Wiring diagram, CB750 F2 (UK model)*

*For US diagram see page 126

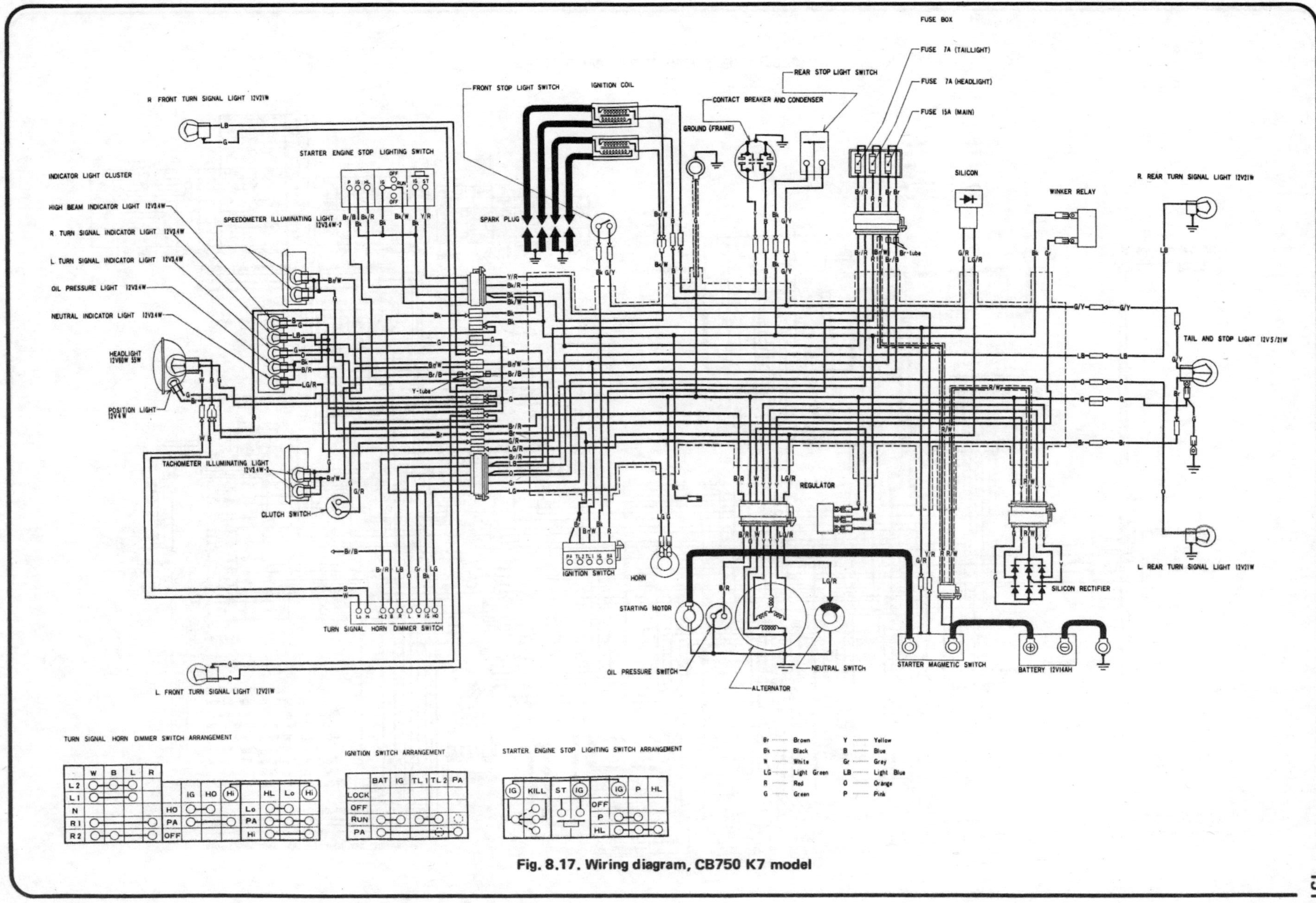

Fig. 8.17. Wiring diagram, CB750 K7 model

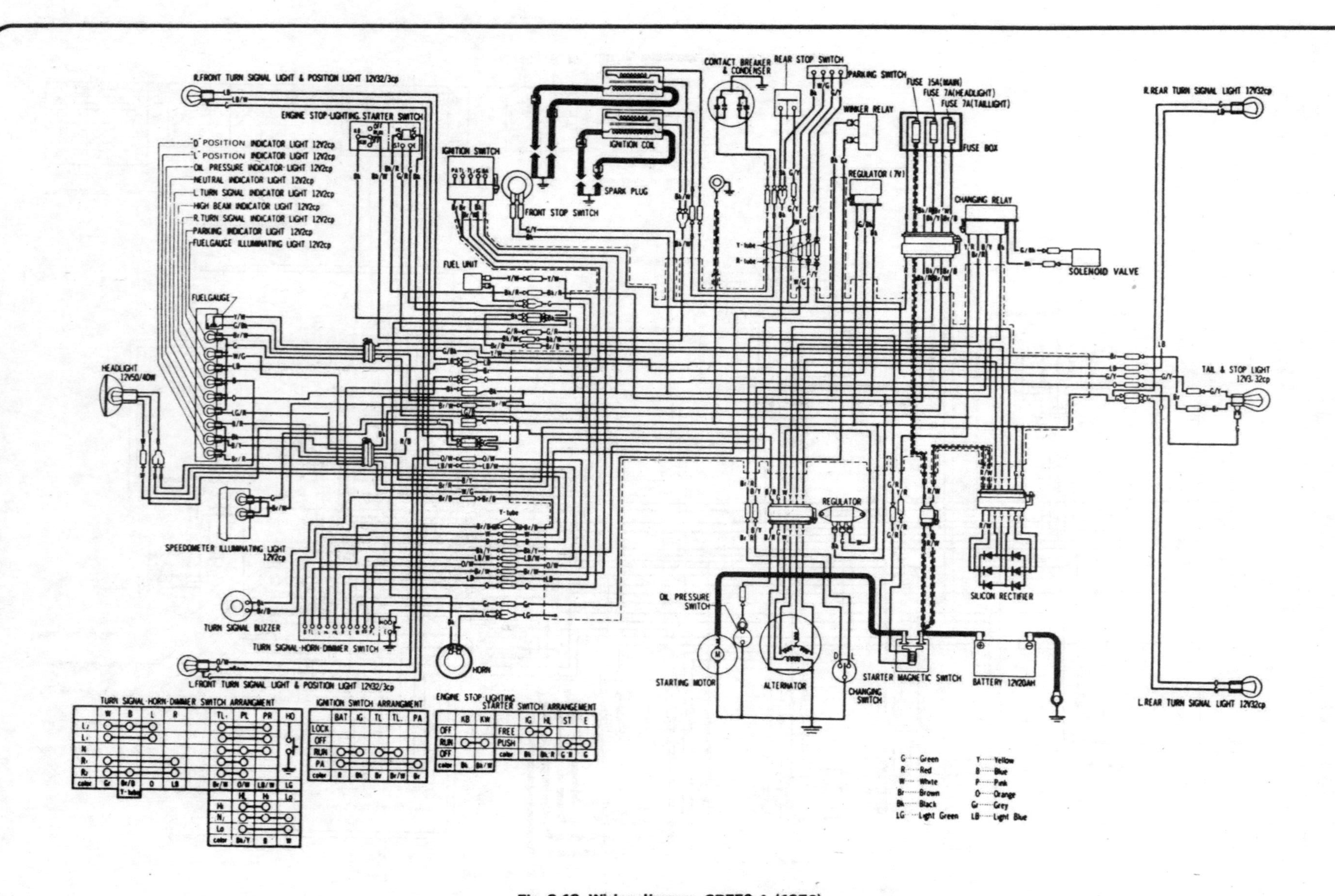

Fig. 8.18. Wiring diagram, CB750 A (1976)

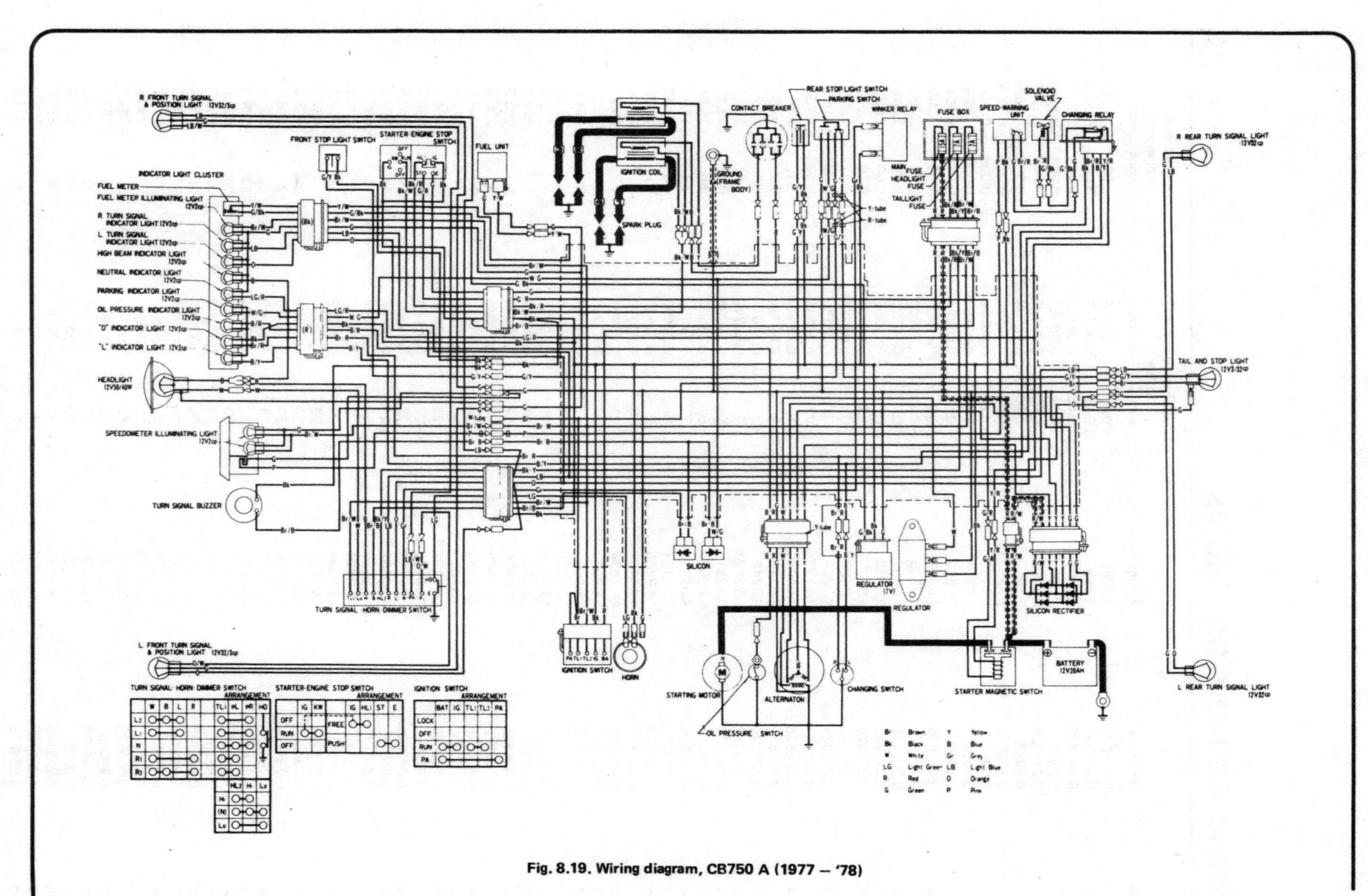

Fig. 8.19. Wiring diagram, CB750 A (1977 – '78)

Metric conversion tables

Inches	Decimals	Millimetres	Millimetres to Inches		Inches to Millimetres	
			mm	Inches	Inches	mm
1/64	0.015625	0.3969	0.01	0.00039	0.001	0.0254
1/32	0.03125	0.7937	0.02	0.00079	0.002	0.0508
3/64	0.046875	1.1906	0.03	0.00118	0.003	0.0762
1/16	0.0625	1.5875	0.04	0.00157	0.004	0.1016
5/64	0.078125	1.9844	0.05	0.00197	0.005	0.1270
3/32	0.09375	2.3812	0.06	0.00236	0.006	0.1524
7/64	0.109375	2.7781	0.07	0.00276	0.007	0.1778
1/8	0.125	3.1750	0.08	0.00315	0.008	0.2032
9/64	0.140625	3.5719	0.09	0.00354	0.009	0.2286
5/32	0.15625	3.9687	0.1	0.00394	0.01	0.254
11/64	0.171875	4.3656	0.2	0.00787	0.02	0.508
3/16	0.1875	4.7625	0.3	0.01181	0.03	0.762
13/64	0.203125	5.1594	0.4	0.01575	0.04	1.016
7/32	0.21875	5.5562	0.5	0.01969	0.05	1.270
15/64	0.234375	5.9531	0.6	0.02362	0.06	1.524
1/4	0.25	6.3500	0.7	0.02756	0.07	1.778
17/64	0.265625	6.7469	0.8	0.03150	0.08	2.032
9/32	0.28125	7.1437	0.9	0.03543	0.09	2.286
19/64	0.296875	7.5406	1	0.03947	0.1	2.54
5/16	0.3125	7.9375	2	0.07874	0.2	5.08
21/64	0.328125	8.3344	3	0.11811	0.3	7.62
11/32	0.34375	8.7312	4	0.15748	0.4	10.16
23/64	0.359375	9.1281	5	0.19685	0.5	12.70
3/8	0.375	9.5250	6	0.23622	0.6	15.24
25/64	0.390625	9.9219	7	0.27559	0.7	17.78
13/32	0.40625	10.3187	8	0.31496	0.8	20.32
27/64	0.421875	10.7156	9	0.35433	0.9	22.86
7/16	0.4375	11.1125	10	0.39370	1	25.4
29/64	0.453125	11.5094	11	0.43307	2	50.8
15/32	0.46875	11.9062	12	0.47244	3	76.2
31/64	0.484375	12.3031	13	0.51181	4	101.6
1/2	0.5	12.7000	14	0.55118	5	127.0
33/64	0.515625	13.0969	15	0.59055	6	152.4
17/32	0.53125	13.4937	16	0.62992	7	177.8
35/64	0.546875	13.8906	17	0.66929	8	203.2
9/16	0.5625	14.2875	18	0.70866	9	228.6
37/64	0.578125	14.6844	19	0.74803	10	254.0
19/32	0.59375	15.0812	20	0.78740	11	279.4
39/64	0.609375	15.4781	21	0.82677	12	304.8
5/8	0.625	15.8750	22	0.86614	13	330.2
41/64	0.640625	16.2719	23	0.90551	14	355.6
21/32	0.65625	16.6687	24	0.94488	15	381.0
43/64	0.671875	17.0656	25	0.98425	16	406.4
11/16	0.6875	17.4625	26	1.02362	17	431.8
45/64	0.703125	17.8594	27	1.06299	18	457.2
23/32	0.71875	18.2562	28	1.10236	19	482.6
47/64	0.734375	18.6531	29	1.14173	20	508.0
3/4	0.75	19.0500	30	1.18110	21	533.4
49/64	0.765625	19.4469	31	1.22047	22	558.8
25/32	0.78125	19.8437	32	1.25984	23	584.2
51/64	0.796875	20.2406	33	1.29921	24	609.6
13/16	0.8125	20.6375	34	1.33858	25	635.0
53/64	0.828125	21.0344	35	1.37795	26	660.4
27/32	0.84375	21.4312	36	1.41732	27	685.8
55/64	0.859375	21.8281	37	1.4567	28	711.2
7/8	0.875	22.2250	38	1.4961	29	736.6
57/64	0.890625	22.6219	39	1.5354	30	762.0
29/32	0.90625	23.0187	40	1.5748	31	787.4
59/64	0.921875	23.4156	41	1.6142	32	812.8
15/16	0.9375	23.8125	42	1.6535	33	838.2
61/64	0.953125	24.2094	43	1.6929	34	863.6
31/32	0.96875	24.6062	44	1.7323	35	889.0
63/64	0.984375	25.0031	45	1.7717	36	914.4

English/American terminology

Because this book has been written in England, British English component names, phrases and spellings have been used throughout. American English usage is quite often different and whereas normally no confusion should occur, a list of equivalent terminology is given below.

English	American	English	American
Air filter	Air cleaner	Number plate	License plate
Alignment (headlamp)	Aim	Output or layshaft	Countershaft
Allen screw/key	Socket screw/wrench	Panniers	Side cases
Anticlockwise	Counterclockwise	Paraffin	Kerosene
Bottom/top gear	Low/high gear	Petrol	Gasoline
Bottom/top yoke	Bottom/top triple clamp	Petrol/fuel tank	Gas tank
Bush	Bushing	Pinking	Pinging
Carburettor	Carburetor	Rear suspension unit	Rear shock absorber
Catch	Latch	Rocker cover	Valve cover
Circlip	Snap ring	Selector	Shifter
Clutch drum	Clutch housing	Self-locking pliers	Vise-grips
Dip switch	Dimmer switch	Side or parking lamp	Parking or auxiliary light
Disulphide	Disulfide	Side or prop stand	Kick stand
Dynamo	DC generator	Silencer	Muffler
Earth	Ground	Spanner	Wrench
End float	End play	Split pin	Cotter pin
Engineer's blue	Machinist's dye	Stanchion	Tube
Exhaust pipe	Header	Sulphuric	Sulfuric
Fault diagnosis	Trouble shooting	Sump	Oil pan
Float chamber	Float bowl	Swinging arm	Swingarm
Footrest	Footpeg	Tab washer	Lock washer
Fuel/petrol tap	Petcock	Top box	Trunk
Gaiter	Boot	Torch	Flashlight
Gearbox	Transmission	Two/four stroke	Two/four cycle
Gearchange	Shift	Tyre	Tire
Gudgeon pin	Wrist/piston pin	Valve collar	Valve retainer
Indicator	Turn signal	Valve collets	Valve cotters
Inlet	Intake	Vice	Vise
Input shaft or mainshaft	Mainshaft	Wheel spindle	Axle
Kickstart	Kickstarter	White spirit	Stoddard solvent
Lower leg	Slider	Windscreen	Windshield
Mudguard	Fender		

Index